Mathematics for Industry

Volume 30

Aims & Scope

The meaning of "Mathematics for Industry" (sometimes abbreviated as MI or MfI) is different from that of "Mathematics in Industry" (or of "Industrial Mathematics"). The latter is restrictive: it tends to be identified with the actual mathematics that specifically arises in the daily management and operation of manufacturing. The former, however, denotes a new research field in mathematics that may serve as a foundation for creating future technologies. This concept was born from the integration and reorganization of pure and applied mathematics in the present day into a fluid and versatile form capable of stimulating awareness of the importance of mathematics in industry, as well as responding to the needs of industrial technologies. The history of this integration and reorganization indicates that this basic idea will someday find increasing utility. Mathematics can be a key technology in modern society.

The series aims to promote this trend by (1) providing comprehensive content on applications of mathematics, especially to industry technologies via various types of scientific research, (2) introducing basic, useful, necessary and crucial knowledge for several applications through concrete subjects, and (3) introducing new research results and developments for applications of mathematics in the real world. These points may provide the basis for opening a new mathematics oriented technological world and even new research fields of mathematics.

More information about this series at http://www.springer.com/series/13254

Patrick van Meurs · Masato Kimura
Hirofumi Notsu
Editors

Mathematical Analysis of Continuum Mechanics and Industrial Applications II

Proceedings of the International Conference CoMFoS16

Springer

Editors
Patrick van Meurs
Kanazawa University
Kanazawa, Ishikawa
Japan

Hirofumi Notsu
Kanazawa University
Kanazawa, Ishikawa
Japan

Masato Kimura
Kanazawa University
Kanazawa, Ishikawa
Japan

ISSN 2198-350X ISSN 2198-3518 (electronic)
Mathematics for Industry
ISBN 978-981-10-6282-7 ISBN 978-981-10-6283-4 (eBook)
https://doi.org/10.1007/978-981-10-6283-4

Library of Congress Control Number: 2017950285

Printed on acid-free paper

This Springer imprint is published by Springer Nature
The registered company is Springer Nature Singapore Pte Ltd.
The registered company address is: 152 Beach Road, #21-01/04 Gateway East, Singapore 189721, Singapore

Preface

The starting point of the conference series CoMFoS "Continuum Mechanics Focusing on Singularities" dates back twenty-one years ago, when a small workshop on mathematical fracture mechanics was organized in Ishikawa prefecture in Japan. The organizers of this workshop were researchers in both mathematics and engineering, who firmly believed the importance of mathematics in further development of engineering and industry, especially in the field of continuum mechanics. These organizers were Kohji Ohtsuka (Hiroshima Kokusai Gakuin University), Tetsuhiko Miyoshi (Yamaguchi University), and Chikayoshi Yatomi (Kanazawa University), who are considered the founding members of the series CoMFoS. Their activity continues today in the research activity group "Mathematical Aspects of Continuum Mechanics (MACM)", which is part of the Japan Society for Industrial and Applied Mathematics (JSIAM).

One of the main examples of the importance of the conference series CoMFoS stems from the rapid growth of computers over the last two decades, which extremely enhanced the computer simulation techniques in engineering and industrial software packages. These techniques often lack a rigorous mathematical foundation, which lead to questionable consistency, accuracy, and efficiency. Improving such properties can only be done by combining the knowledge from researchers from industry, engineering, and mathematics, and that is exactly what the series CoMFoS intends to do.

The international conference CoMFoS16 "Mathematical Analysis of Continuum Mechanics and Industrial Applications II" was held at Kyushu University, Fukuoka, Japan, from October 22 to 24, 2016, which was supported and organized by Institute of Mathematics for Industry (IMI), Kyushu University, as part of the project "Workshop of the Joint Research Projects", and was co-organized by MACM. It was the sequel of previous year's CoMFoS15 as the 16th meeting of CoMFoS. It attracted a wide variety of researchers not only from mathematics and engineering but also from fields such as rheology, nano-science, material science, and seismology. It achieved a great success under the financial and operational supports of IMI, Kyushu University.

It is our great pleasure to present the proceeding of CoMFoS16 as part of the series of Mathematics for Industry, following up on the proceedings of CoMFoS15, which has recently been published in the same series. This book consists of fifteen peer-reviewed research papers and surveys, which are divided into the following six chapters: Fracture Mechanics, Shape Optimization, Modeling of Earthquakes, Material Structure, Interface Dynamics, and Complex Systems. The wide variety of the topics contained in this proceedings reflects the increasing presence and importance of the mathematical modeling and analysis in the present and future technologies and sciences.

Kanazawa, Japan

Masato Kimura
On behalf of the Organizing Committee of CoMFoS16

Contents

Part I
Fracture Mechanics

Mathematical Modeling of the Desiccation Cracking

Sayako Hirobe

Abstract Desiccation cracks have a net-like appearance and tessellate the drying surface of the materials into polygonal cells with the typical size. The basic features of the crack pattern and the pattern formation process are conserved regardless of the choice of the materials. This implies the existence of the common governing mechanism behind the pattern formation in desiccation cracking. We propose the coupled model of desiccation, deformation, and fracture for the desiccation crack phenomenon in the framework of the continuum mechanics. By using this coupled model and the appropriate numerical analysis methods, the typical geometry and the typical length scale of the desiccation crack pattern are reproduced in the complete homogeneous field without any artificial length scale. These results indicate that the proposed coupled model captures the fundamental mechanism for the pattern formation in desiccation cracking.

1 Introduction

Desiccation cracks are widely observed in nature such as desserts, mudflats, paddy fields, and graze on ceramics. Generally, these cracks have a net-like appearance and tessellate the drying surfaces of the materials into polygonal cells with the typical size. This curious crack pattern attracts many researchers' attention in the last century in both experimental and numerical approaches.

The experimental researches have been performed with various materials and conditions to investigate the relationship between the cell geometry (i.e., the size and the shape of the cells) and the experimental conditions [3, 6–8, 12, 14]. These researches show that the typical cell size has a strong dependence on the layer thickness. Furthermore, some of these researches intensively study the process of the crack pattern formation. They pointed out that the crack patterns are formed in a hierarchical cell tessellation process. The larger cells are successively tessellated by the new cracks

S. Hirobe (✉)
Keio University, Hiyoshi 3-14-1, Kohoku-ku, Yokohama 223-8522, Japan
e-mail: s.hirobe@keio.jp

P. van Meurs et al. (eds.), *Mathematical Analysis of Continuum Mechanics and Industrial Applications II*, Mathematics for Industry 30,
https://doi.org/10.1007/978-981-10-6283-4_1

until the cell size reaches to the typical cell size observed in the final pattern. While some varieties can be observed in the shape of the cracks depending on the materials, these basic features of the crack pattern and the pattern formation process are conserved regardless of the choice of the materials. This implies the existence of the common governing mechanism behind the pattern formation in desiccation cracking.

The previous numerical approaches have proposed some mathematical model and numerical analysis methods for the analysis of the desiccation crack phenomenon [1, 2, 10, 11, 13, 15, 16]. However, these models and methods assume the homogeneous water distribution and the heterogeneity in the material properties. This assumption corresponds to the introduction of the artificial length scale in the models. Therefore, the achievement of these previous numerical approaches is limited to the reproduction of the final crack pattern observed in the thin layer (the assumption of the homogeneous water distribution is applicable only for the thin layer). Thus, they cannot propose the effective description of the fundamental mechanism for the emergence of the characteristic crack pattern in the desiccation crack phenomenon.

In this research, a mathematical model for the desiccation crack phenomenon is proposed in the framework of continuum mechanics. The desiccation crack phenomenon is modeled as the coupling of three physics: desiccation, deformation, and fracture. By using this coupled model, we show that the typical geometry and the typical length scale of the desiccation crack pattern can be reproduced in the completely homogeneous field without any artificial length scale.

2 Drying Experiment

We performed the drying test with calcium carbonate slurry to observe the basic features of the desiccation crack phenomenon. The calcium carbonate slurry (the volumetric water content of the slurry was 72.0%) was poured into the rectangular acrylic container (100 mm width and height, and 50 mm depth). We prepared three specimens with different thickness: 5, 10, and 20 mm. These specimens were dried in the air at the temperature of 20 °C and a relative humidity of 50% until the specimens were completely dried. We observed the crack propagation process and the weight reduction of the specimens every 10 min.

Figure 1 shows the crack pattern formed on the top surface of the specimens with different thickness. For each specimen, the polygonal cells framed by the cracks are formed, and the size of the cells is kept almost constant except for the boundary cells affected by the container walls. Also, we can observe the increasing tendency of the average cell size with the increase of the specimen thickness. These results indicate that the polygonal cells have the typical sizes, and this size changes systematically depending on the thickness of the specimen.

Figure 2 shows the crack propagation process on the top surface of the specimen with 10 mm thickness. At the initial stage of the desiccation cracking, some long cracks tessellate the specimen surface and form a largest structure of the cells. Then,

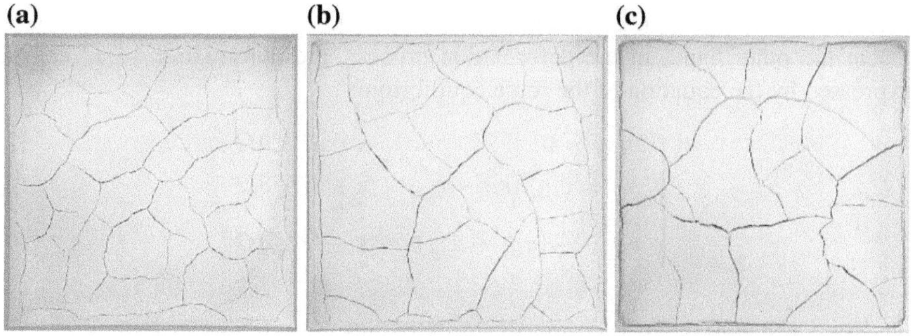

Fig. 1 The crack pattern formed on the top surface of the specimens with different thickness. **a** 5 mm, **b** 10 mm, and **c** 20 mm

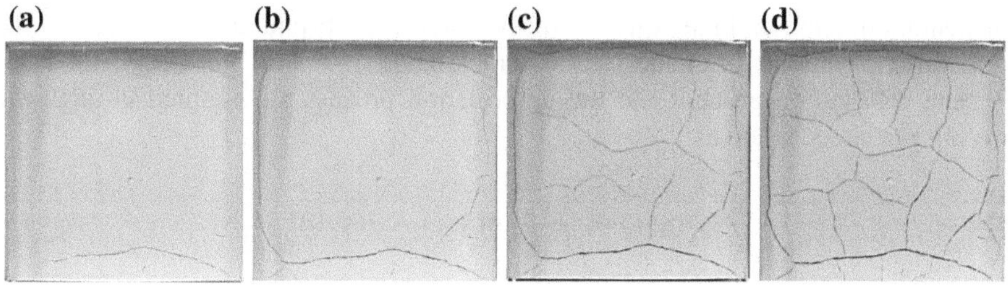

Fig. 2 The crack propagation process on the top surface of the specimen with 10 mm thickness

relatively short cracks grow to tessellate the larger cells. This hierarchical sequence of the cell tessellation is commonly observed in the desiccation cracking.

3 Coupled Model for the Desiccation Crack Phenomenon

3.1 Mathematical Model

Consider a the homogeneous and isotropic linearly elastic body Ω with the external boundary $\partial\Omega$. The external boundary of Ω can be divided into $\partial\Omega^{(1)}$ and $\partial\Omega^{(2)}$: $\partial\Omega = \partial\Omega^{(1)} \cup \partial\Omega^{(2)}$, $\partial\Omega^{(1)} \cap \partial\Omega^{(2)} = \emptyset$. In the case of the drying experiment shown in Sect. 2, $\partial\Omega^{(1)}$ corresponds to the specimen surfaces adjoining the container walls and $\partial\Omega^{(2)}$ corresponds to the evaporation surface which is opening to the air.

In the desiccation process, the water movement in Ω is simplified to the diffusion equation in terms of the volumetric water content $\theta(\mathbf{x}, t)$:

$$\dot{\theta} = D\nabla^2\theta \quad \mathbf{x} \in \Omega \setminus \partial\Omega \tag{1}$$

where D is the diffusion coefficient.

On the other hand, in the deformation process, the deformation of Ω can be expressed by the equation of the force equilibrium:

$$\begin{cases} \sigma_{ij,j} = 0 & \mathbf{x} \in \Omega \setminus \partial\Omega & (2) \\ \sigma_{ij} = c_{ijkl}\left(\varepsilon_{kl} - \varepsilon^{s}_{kl}\right) & \mathbf{x} \in \Omega \setminus \partial\Omega & (3) \\ \varepsilon_{ij} = \dfrac{1}{2}\left(u_{i,j} + u_{j,i}\right) & \mathbf{x} \in \Omega \setminus \partial\Omega & (4) \end{cases}$$

where σ_{ij} is a stress tensor, c_{ijkl} is an elasticity tensor, ε_{ij} is a total strain tensor, ε^{s}_{ij} is a shrinkage strain tensor, and u_i is a displacement vector. In the case of the drying shrinkage, the drying shrinkage strain ε^{s}_{ij} is inelastic strain resulting from the volume reduction, and therefore, it does not contribute to the generation of the stress. The diffusion equation and the equation of the force equilibrium are introduced as the governing equations for the desiccation cracking.

The desiccation process and the deformation process are coupled through the drying shrinkage strain as

$$\varepsilon^{v}(\mathbf{x}, t) = \frac{1}{\lambda}\frac{\rho_w}{\rho_b}\{\theta(\mathbf{x}, t) - \theta(\mathbf{x}, 0)\} \tag{5}$$

$$\varepsilon^{s}_{ij} = \frac{1}{3}\varepsilon^{v}\delta_{ij} \tag{6}$$

where ε^{v} is a volumetric drying shrinkage strain, λ is a moisture shrinkage coefficient of the powder, ρ_w is a mass density of the water, ρ_b is a dry bulk density of the powder, and δ_{ij} is the Kronecker's delta. In Eq. (5), we considered the isotropy and the homogeneity of Ω.

When the stress defined exceeds the tensile strength, the crack surfaces Γ are formed in Ω ($\Gamma \in \Omega$). These crack surfaces Γ work as the internal evaporation surfaces in the desiccation process and the internal traction-free surfaces in the deformation process. Therefore, the initial boundary value problem for the desiccation field with crack surfaces and the boundary value problem for the deformation field with crack surfaces are given as

$$\begin{cases} \dot{\theta} = D\nabla^2\theta & \mathbf{x} \in \Omega \setminus (\partial\Omega \cup \Gamma) & (7a) \\ \theta(\mathbf{x}, 0) = \theta_0(\mathbf{x}) & \mathbf{x} \in \Omega \setminus (\partial\Omega \cup \Gamma) & (7b) \\ D\dfrac{\partial\theta}{\partial\mathbf{n}} = 0 & \mathbf{x} \text{ on } \partial\Omega^{(1)} & (7c) \\ D\dfrac{\partial\theta}{\partial\mathbf{n}} = -q^{\Omega}(\mathbf{x}, t) & \mathbf{x} \text{ on } \partial\Omega^{(2)} & (7d) \\ D\dfrac{\partial\theta}{\partial\mathbf{n}} = -q^{\Gamma}(\mathbf{x}, t) & \mathbf{x} \text{ on } \Gamma & (7e) \end{cases}$$

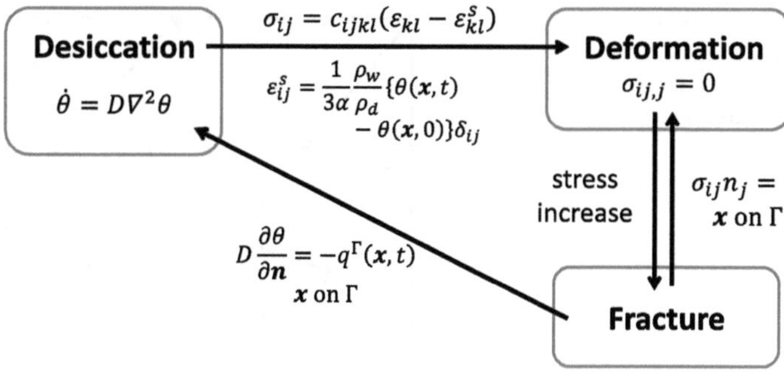

Fig. 3 The structure of the coupled model

$$\begin{cases} \sigma_{ij,j} = 0 & \mathbf{x} \in \Omega \setminus (\partial\Omega \cup \Gamma) & \text{(8a)} \\ \sigma_{ij} = c_{ijkl}\left(\varepsilon_{kl} - \varepsilon^s_{kl}\right) & \mathbf{x} \in \Omega \setminus (\partial\Omega \cup \Gamma) & \text{(8b)} \\ \varepsilon_{ij} = \frac{1}{2}\left(u_{i,j} + u_{j,i}\right) & \mathbf{x} \in \Omega \setminus (\partial\Omega \cup \Gamma) & \text{(8c)} \\ u_i = 0 & \mathbf{x} \text{ on } \partial\Omega^{(1)} & \text{(8d)} \\ \sigma_{ij}n_j = 0 & \mathbf{x} \text{ on } \partial\Omega^{(2)} & \text{(8e)} \\ \sigma_{ij}n_j = 0 & \mathbf{x} \text{ on } \Gamma & \text{(8f)} \end{cases}$$

where $\theta_0(\mathbf{x})$ is the initial distribution of the volumetric water content, $\mathbf{n}$ is a unit normal vector for $\partial\Omega^{(2)}$, and Γ, q^Ω is the evaporation speed prescribed on $\partial\Omega^{(2)}$, and q^Γ is the evaporation speed prescribed on Γ. Here, the specimen surfaces adjoining the container wall ($\partial\Omega^{(1)}$) is constrained by the friction and the adhesion between the specimen surfaces and the container walls. Therefore, the desiccation crack phenomenon can be modeled as the coupling of the initial boundary value problem for the desiccation process (7) and the boundary value problem for the deformation process (8).

The structure of this coupled model is illustrated as Fig. 3. The desiccation and the deformation are coupled through the drying shrinkage strain ε^s_{ij}, and the excessive stress due to deformation causes the fracture. The effect of the fracture on the desiccation and the deformation is embedded as the internal evaporation surfaces and the internal traction-free surfaces. As seen in this figure, the direct interaction exists only between the deformation and the fracture. Other relationships (between the desiccation and the deformation, between the fracture and the desiccation) are not an interaction but one-way actions. To solve this coupled problem numerically, (i) the strong coupling analysis of the deformation problem and the fracture problem, and (ii) the weak coupling analysis of the desiccation problem and the deformation–fracture problem are performed. This proper use of the coupling analysis methods is determined based on the interaction among three physics.

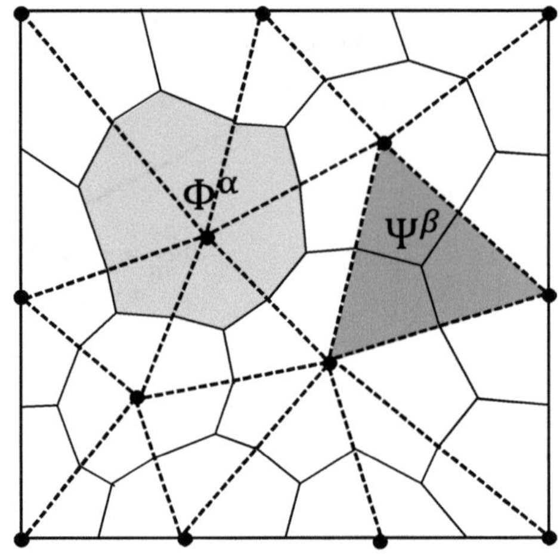

Fig. 4 The pair of conjugate geometries (the Voronoi tessellations $\{\Phi^\alpha\}$ and the Delaunay tessellations $\{\Psi^\beta\}$)

3.2 *Numerical Analysis Methods*

For the strong coupling analysis of the deformation and the fracture, we use Particle Discretization Scheme Finite Element Method (PDS-FEM) [4, 5, 9, 17]. PDS-FEM is a fracture analysis method which can treat the deformation and fracture problem seamlessly. PDS-FEM applies the particle discretization to the field variables by using the discontinuous and non-overlapping shape functions on the pair of the conjugate geometries (the Voronoi tessellations and the Delaunay tessellations shown in Fig. 4). The Delaunay tessellations correspond to the linear triangular (tetrahedral) elements used in the ordinary FEM. The shape functions defined on the Voronoi tessellations $\{\Phi^\alpha\}$ and the Delaunay tessellations $\{\Psi^\beta\}$ are

$$\phi^\alpha(\mathbf{x}) = \begin{cases} 1 \ (\mathbf{x} \in \Phi^\alpha) \\ 0 \ (\mathbf{x} \notin \Phi^\alpha), \end{cases} \tag{9}$$

$$\psi^\beta(\mathbf{x}) = \begin{cases} 1 \ (\mathbf{x} \in \Psi^\beta) \\ 0 \ (\mathbf{x} \notin \Psi^\beta). \end{cases} \tag{10}$$

According to these shape functions, the discontinuity in the displacement field due to fracture is easily represented.

The displacement field is discretized by the Voronoi tessellations $\{\Phi^\alpha\}$ as

$$u_i(\mathbf{x}) = \sum_{\alpha=1}^{N} u_i^\alpha \phi^\alpha(\mathbf{x}) \tag{11}$$

where N is the number of the Voronoi tessellations (corresponding to the number of nodes in FEM). This discretized displacement field corresponds to the translational motion of the rigid body particles. On the other hand, the field variables related to the spatial derivatives of the displacement u_i (i.e., the strain and the stress) are discretized by the Delaunay tessellations $\{\Psi^\beta\}$ as

$$\varepsilon_{ij}(\mathbf{x}) = \sum_{\beta=1}^{M} \varepsilon_{ij}^{\beta} \psi^{\beta}(\mathbf{x}) \tag{12}$$

$$\sigma_{ij}(\mathbf{x}) = \sum_{\beta=1}^{M} \sigma_{ij}^{\beta} \psi^{\beta}(\mathbf{x}) \tag{13}$$

where M is the number of Delaunay tessellations. Thus, the strain and the stress field are averaged over the Delaunay tessellations. The Delaunay tessellation is the only choice for the appropriate averaging domain for the strain and the stress filed.

Applying the discretization shown in Eqs. (11)–(13) to the set of equations of the deformation process Eq. (8), the discretized governing equation for the deformation process is given as

$$\sum_{\gamma=1}^{N} \mathbf{K}^{\alpha\gamma} \mathbf{u}^{\gamma} = \mathbf{F}^{\alpha} \tag{14}$$

where $K_{ik}^{\alpha\gamma}$ is a stiffness matrix and $\mathbf{F}^{\alpha}$ is the external force vector on the α-th Voronoi tessellation; the detailed explanation for this discretization scheme is seen in [4, 5, 9, 17]. This stiffness matrix is identical to that of the ordinary FEM using the triangular (tetrahedral) elements with a linear interpolation function. The effect of the drying shrinkage strain ε_{ij}^{s} is embedded in the external force vector $\mathbf{F}^{\alpha}$.

In PDS-FEM, the crack surfaces are defined on the boundary of the Voronoi tessellations, and the fracture is expressed by the loss of the interaction between the Voronoi tessellations. The interaction between Voronoi tessellations (the interaction between the rigid body particles) corresponds to the stiffness matrix; therefore, the loss of the interaction is expressed by the appropriate change of the stiffness matrix $K_{ik}^{\alpha\gamma}$. In this treatment of the fracture, the nodes are not placed on the traction-free crack surfaces. Thus, the cracks are implemented as the blunt cracks. This geometric implementation of the crack surfaces for the deformation problem is inherited to the analysis of the desiccation problem. The rigorous formulation of PDS-FEM and the treatment of the fracture should be referred to [4, 5, 9, 17].

In this research, the desiccation problem is solved by FEM with linear tetrahedral elements. The geometry of the tetrahedral elements used in FEM analysis for the desiccation problem is identical to the geometry of the Delaunay tessellations used in the PDS-FEM analysis. The FEM formulation for the set of equations of the desiccation process (Eq. 7) with linear interpolation function is given as

$$\sum_{\gamma=1}^{N} \mathbf{M}^{\alpha\gamma} \dot{\theta}^{\gamma} + \sum_{\gamma=1}^{N} \mathbf{A}^{\alpha\gamma} \theta^{\gamma} = \mathbf{Q}^{\alpha}. \tag{15}$$

The explicit form of $\mathbf{M}^{\alpha\gamma}$, $\mathbf{A}^{\alpha\gamma}$, and $\mathbf{Q}^{\alpha}$ can be easily obtained through the finite element formulation.

As mentioned above, the crack surfaces Γ work as the newly created evaporation surfaces. Since the crack surfaces are implemented as the blunt cracks in the discretized form, the Neumann boundary condition (7e) prescribed on the crack surfaces Γ should be divided into (i) the anisotropic diffusion coefficient in a fractured element and (ii) the prescribed evaporation speed at the nodes adjacent to the crack surfaces. The anisotropic diffusion coefficient is expressed as the elimination of the water flux in the normal direction on the crack surfaces. This anisotropic diffusion coefficient is introduced to the $\mathbf{A}^{\alpha\gamma}$ matrix in Eq. (15).

Thus, the crack generation (i.e., the emergence of the strong discontinuity in the continuous field) is expressed as the change in the matrices ($\mathbf{K}^{\alpha\gamma}$ in the deformation problem and $\mathbf{A}^{\alpha\gamma}$ in the desiccation problem) and the nodal force (the prescribed evaporation speed at the nodes adjacent to the crack surfaces in the desiccation problem) in the discretized coupled model. The FEM analysis for the desiccation and the PDS-FEM analysis for the deformation and the fracture are weakly coupled.

4 Numerical Analysis

Based on the proposed coupled model and numerical analysis methods, we performed three-dimensional numerical analysis to reproduce the crack patterns and the crack propagation process observed in the drying test with calcium carbonate slurry. We prepare the rectangular parallelepiped analysis model with the unstructured tetrahedral mesh; the width and the hight are set as 100 mm and the thickness is set as 5, 10, and 20 mm according to the drying test. The mesh sizes for each analysis model are shown in Table 1. The nodal displacement is constrained on the sides and the bottom surface of the analysis model. The evaporation speed is prescribed on the top surface of the analysis model and the crack surfaces. The parameters used for this analysis are shown in Table 2. The initial volumetric water content θ_0, the final volumetric water content θ_{fin}, and the evaporation speed on the crack surfaces q^{Γ} are determined from the drying experiment with calcium carbonate slurry in Sect. 2. Since the evaporation speed on the crack surfaces q^{Γ} is considered as slower than the evaporation speed on the top surface q^{Ω}, q^{Γ} is set as 50% of q^{Ω}.

The FEM analysis for the desiccation problem is proceeded with the constant time step $\Delta t = 0.1$ h by using the backward Euler method. In each time step, the quasi-static fracture analysis by PDS-FEM is performed after the FEM analysis. When the maximum traction among all elements reaches to 97% of the tensile strength t_c, the time step is reduced to $\Delta t = 0.01$ h to capture the effect of the cracks promptly. The

Table 1 Mesh sizes of each analysis model

Model size (mm)	Number of elements	Number of nodes
100 × 100 × 5	160,732	32,802
100 × 100 × 10	326,368	60,215
100 × 100 × 20	325,931	58,129

Table 2 The parameters used in the numerical analysis

Dry bulk density ρ_b	8.0×10^2 kg/m^3
Mass density of water ρ_w	1.0×10^3 kg/m^3
Initial volumetric water content θ_0	0.560
Final volumetric water content θ_{fin}	0.204
Evaporation speed on the top surface q^{Ω}	8.8×10^{-5} m/h
Evaporation speed on the crack surfaces q^{Γ}	4.4×10^{-5} m/h
Moisture shrinkage coefficient λ	0.69
Moisture diffusion coefficient D	3.6×10^{-6} m^2/h
Poisson's ratio ν	0.30
Young's modulus E	5.0 MPa
Tensile strength t_c	1.6 MPa

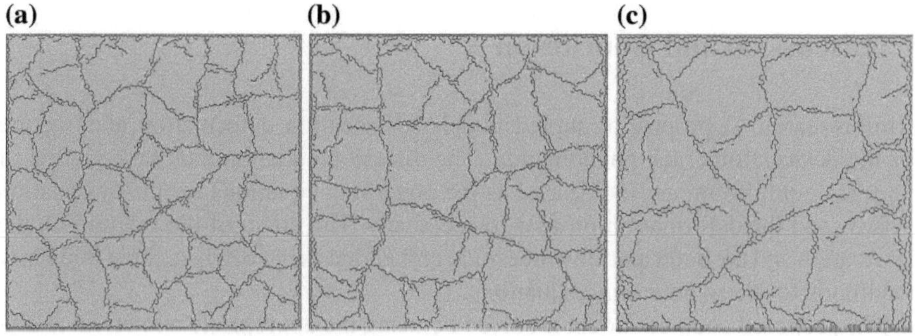

Fig. 5 The final crack pattern formed on the top surface of the analysis model with different thickness. **a** 5 mm, **b** 10 mm, and **c** 20 mm

numerical analysis proceeds until the average volumetric water content reaches to the final volumetric water content θ_{fin}.

Figure 5 shows the final crack pattern formed on the top surface of the analysis model with different thickness. The desiccation time required for reaching the final volumetric water content θ_{fin} (the final crack patterns) is 26.59, 44.92, and 73.85 h for 5, 10, and 20 mm thickness analysis models, respectively. As seen in Fig. 5, the net-like cracks form the polygonal cells with the typical size. This typical cell size increases as the increase of the model thickness. Most of the cracks is vertical to the

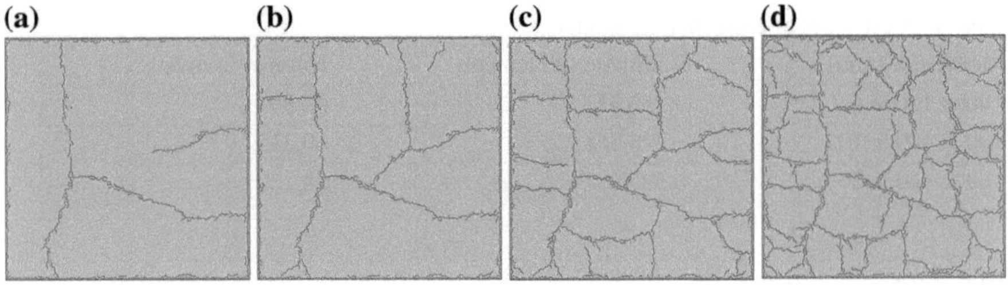

Fig. 6 The crack propagation process on the top surface of the analysis model with 10 mm thickness

bottom surface and reached to the bottom surface. These features of the desiccation crack pattern coincide with the observation in the drying experiment with calcium carbonate slurry.

Figure 6 shows the crack propagation process on the top surface of the analysis model with 10 mm thickness. At the initial stage of the crack pattern formation, some cracks extend long and form the largest structure of the cells. Then, some relatively short cracks tessellate the larger cells. This hierarchical sequence of the cell tessellation is also observed in the drying test with calcium carbonate slurry.

5 Discussion and Conclusion

In this research, we propose a coupled model of desiccation, deformation, and fracture for the desiccation crack phenomenon. The diffusion equation and the equation of the force equilibrium are introduced as the governing equations for the desiccation process and the deformation process, respectively. The effect of the fracture on the desiccation and the deformation process are embedded as the newly created boundary conditions for these governing equations.

The proposed coupled model is solved by using the appropriate numerical analysis methods: FEM and PDS-FEM. The FEM analysis for the desiccation and the PDS-FEM analysis for the deformation and the fracture are weakly coupled. In spite of the assumption of the completely homogeneous filed, the results of the numerical analysis show the satisfactory agreement with the results of drying test in terms of the cell geometry, the crack propagation process, and the increasing tendency of the typical cell size depending on the specimen thickness. These results indicate that the proposed coupled model captures the fundamental mechanism for the pattern formation in desiccation cracking (i.e., the fundamental mechanism for the emergence of the typical length scale from the completely homogeneous field).

Each governing equation for the desiccation cracking (i.e., the diffusion equation and the equation of the force equilibrium) does not have any length scale. However, the coupling of these governing equations through the fracture can produce the length scale from the completely homogeneous field. This implies that the coupling of the equations without length scale is the governing mechanism for the emergence of the typical length scale in the desiccation crack phenomenon.

References

1. Amarasiri, A.L., Kodikara, J.K., Costa, S.: Int. J. Numer. Anal. Methods Geomech. **35**, 82–96 (2011)
2. Deng, G., Shen, Z.: Geomech. Geoeng. Int. J. **1**, 27–41 (2006)
3. Groisman, A., Kaplan, E.: Europhysics. Lett. **25**, 415–420 (1994)
4. Hori, M., Oguni, K., Sakaguchi, H.: J. Mech. Phys. Solids **53**, 681–703 (2005)
5. Kame, N., Saito, S., Oguni, K.: Geophys. J. Int. **173**, 309–314 (2008)
6. Kindle, E.M.: J. Geol. **25**, 135–144 (1917)
7. Kitsunezaki, S.: Phys. Rev. E **87**, 052805 (2013)
8. Nahlawi, H., Kodikara, J.K.: Geotech. Geol. Eng. **24**, 1641–1664 (2006)
9. Oguni, K., Wijerathne, M.L.L., Okinaka, T., Hori, M.: Mech. Mater. **41**, 1242–1252 (2009)
10. Peron, H., Delenne, J., Laloui, L., El Youssoufi, M.: Comput. Geotech. **36**, 61–69 (2008)
11. Sánchez, M., Manzoli, O.L., Guimarães, L.J.N.: Comput. Geotech. **62**, 27–39 (2014)
12. Shorlin, K.A., de Bruyn, J.R., Graham, M., Morris, S.W.: Phys. Rev. E **61**, 6950 (2000)
13. Sima, J., Jiang, M., Zhou, C.: Comput. Geptech. **56**, 168–180 (2014)
14. Tang, C., Shi, B., Liu, C., Zhao, L., Wang, B.: Eng. Geol. **101**, 204–217 (2008)
15. Trabelsi, H., Jamei, M., Zenzri, H., Olivella, S.: Int. J. Numer. Anal. Methods Geomech. **36**, 1410–1433 (2012)
16. Vogel, H.J., Hoffmann, H., Leopold, A., Roth, K.: Geoderma **125**, 213–223 (2005)
17. Wijerathne, M.L.L., Oguni, K., Hori, M.: Int. J. Numer. Methods Eng. **80**, 46–73 (2009)

Fatigue Crack Growth Analysis of an Interfacial Crack in Heterogonous Material Using XIGA

Indra Vir Singh and Gagandeep Bhardwaj

Abstract In the present work, the fatigue crack growth analysis of an interfacial cracked plate has been performed by extended isogeometric analysis (XIGA). In isogeometric analysis (IGA), non-uniform rational B-splines (NURBS) are employed for defining the geometry as well as the solution. In XIGA, the merits of isogeometric analysis and extended finite element method are combined together for analyzing the cracked geometries. The crack faces are modeled by discontinuous Heaviside jump function, whereas the singularity in the stress field at the crack tip is modeled by crack-tip enrichment functions. The values of stress intensity factors (SIFs) for the interface cracks are evaluated by XIGA and XFEM. Paris law is employed for computing the fatigue life of an interfacial cracked plate.

1 Introduction

Layered materials/coatings are widely used for protecting the components/structures from corrosion and erosion apart from enhancing the strength and hardness. The structural performance of layered materials depends on the mechanical properties and fracture behavior of the interface. Bilayered materials aim to reduce the residual stresses and avoid debonding at the interface. Therefore, the study of bi-material interfacial cracks becomes important for the design of components and structures. These structures/components have got many challenges due to complexity in accurately evaluating the fracture parameters at the interface. The values of these parameters greatly depend on the internal structure of the material, e.g., the stress intensity factor at the crack tip is affected by the presence of flaws in the vicinity of the crack tip.

I.V. Singh (✉)
Department of Mechanical and Industrial Engineering,
Indian Institute of Technology Roorkee, Roorkee, Uttarakhand, India
e-mail: indrafme@iitr.ac.in

G. Bhardwaj
Mechanical Engineering Department, Thapar University, Patiala, Punjab, India
e-mail: gagandeep.med@thapar.edu

P. van Meurs et al. (eds.), *Mathematical Analysis of Continuum Mechanics and Industrial Applications II*, Mathematics for Industry 30,
https://doi.org/10.1007/978-981-10-6283-4_2

Numerical methods are the only choice for an accurate evaluation of the fracture parameters. Over the years, several numerical methods have been developed for evaluating the fracture parameters. Nearly all popular numerical methods discretize a given domain into sub-domains, and then a finite numbers of equations are developed over each domain. The mesh-based methods like FEM, BEM, and FVM can easily and accurately analyze the problems with fixed domain/geometry, but they face a severe challenge in solving the problem involving domain change. To overcome this problem, numerical methods such as element-free Galerkin method [3], boundary node method [14], reproducing kernel particle method [10], cracking particle method [12], and extended finite element method [2] have been developed in past two decades. In these methods, the approximation of geometry introduces some error in the solution as different basis functions are employed for defining the geometry and solution. To cope up with this issue, isogeometric analysis (IGA) has been emerged out as an alternative. In IGA, same basis functions i.e., non-uniform rational B-splines (NURBS) are employed for approximating the geometry as well as the solution. The IGA with partition of unity (PU) enrichment is named as extended isogeometric analysis (XIGA).

2 Isogeometric Analysis

The concept of IGA is introduced by [9] to integrate the computer aided design (CAD) geometry with finite element analysis (FEA). The main aim of IGA is the use of same data set for both design and analysis. In recent times, it gained a lot of attention of the research community due to its accuracy and enhanced capability over standard FEA in analyzing the problems involving complex domains. The NURBS basis functions are employed to approximate the geometry and solution in the physical space. Its name ("isogeometric") is due to fact that the use of NURBS leads to an exact geometric description of the domain, whereas in standard finite element analysis, it is necessary to approximate the geometry by finite elements. The geometry in IGA is modeled accurately even at the coarse mesh, which is difficult to achieve by standard Lagrange basis function in FEA. The details of the B-spline, knot vector, basis function, and NURBS are provided below.

2.1 B-Splines

B-splines are built from piecewise polynomial functions i.e., from a set of polynomial functions. B-spline consists of n-polynomial basis functions of order p.

2.2 Knot Vectors

Knot vectors are used to define the piecewise polynomial basis functions. A knot vector (Ξ) is defined by a set of coordinates in the parametric space $\xi \in [0, 1]$, which provides the information about the subinterval i.e., element. A knot vector (Ξ) in one dimension is defined by a nondecreasing set of coordinates in the parametric space as [4, 7, 13].

$$\Xi = \left\{\xi_1, \xi_2, \ldots\ldots\ldots \xi_{n+p+1}\right\}, \text{with} \quad \xi_i \in R \quad \text{and} \quad \xi_i \leq \xi_{i+1} \tag{1}$$

where, ξ_i and n represent a knot and number of univariate spline basis functions defined on the knot vector, respectively, and i is the knot index whose value varies from $i = 1, 2, 3, \ldots\ldots\ldots, n + p + 1$.

2.3 B-Spline Basis Functions

The B-spline functions are defined recursively starting with the piecewise constants ($p = 0$)

$$N_{i,0}(\xi) = \begin{cases} 1 & \xi_i \leq \xi \leq \xi_{i+1} \\ 0 & otherwise \end{cases} \quad \text{for} \quad p = 0 \tag{2a}$$

$$N_{i,p}(\xi) = \frac{\xi - \xi_i}{\xi_{i+p} - \xi_i} N_{i,p-1}(\xi) + \frac{\xi_{i+p+1} - \xi}{\xi_{i+p+1} - \xi_{i+1}} N_{i+1,p-1}(\xi) \quad \text{for} \quad p \geq 1 \tag{2b}$$

2.4 Non-uniform Rational B-Splines

Non-uniform rational B-splines (NURBS) are piecewise rational polynomials constructed from B-splines. The term nonuniform refers to the use of nonuniform knot vectors. In general, the NURBS are defined as,

$$R_{i,p}(\xi) = \frac{w_i N_{i,p}(\xi)}{W(\xi)} = \frac{w_i N_{i,p}(\xi)}{\sum\limits_{i=0}^{n} w_i N_{i,p}(\xi)} \tag{3}$$

where, $R_{i,p}(\xi)$ defines NURBS, w_i refers to weights associated with the ith control point, and $N_{i,p}(\xi)$ defines B-spline basis functions of order p.

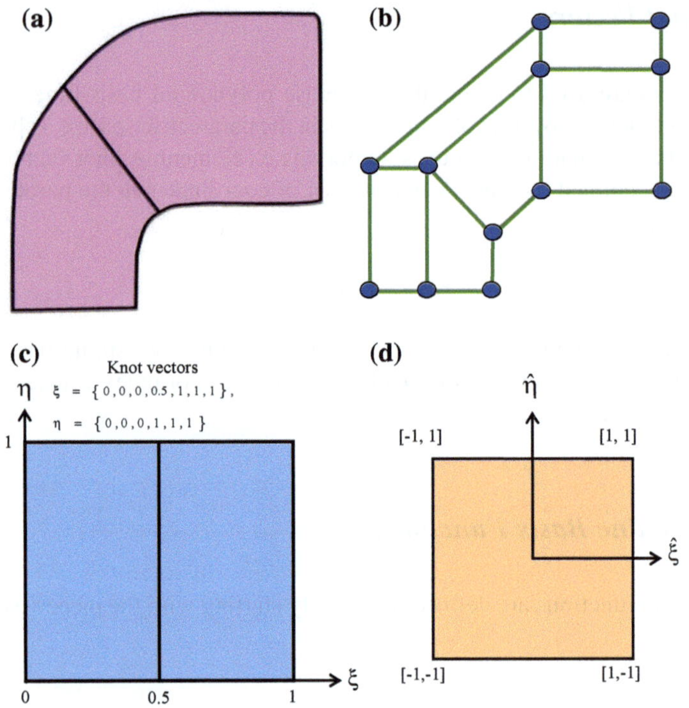

Fig. 1 **a** Physical space. **b** Control mesh. **c** Parametric space. **d** Parent element

2.5 Spaces and Mappings

In isogeometric analysis, different spaces are employed. These spaces are physical space, control mesh, parameter space, index space, and parent element. In physical space, the actual geometry is represented by a linear combination of the basis functions and the control points. The basis functions do not interpolate the control points except at the end. The physical mesh can be divided into elements by knot spans. The physical space is represented in Fig. 1a. The control mesh (control net) is defined by the control points. The control mesh interpolates all control points. It controls the geometry but does not coincide with the physical mesh. The control variables are the degrees of freedom and are located at the control points. A typical control mesh is shown in Fig. 1b. The parametric space is formed by the knot spans. The elements are defined in this space as nonzero knot spans $\left[\xi_i, \xi_{i+1}\right] \times \left[\eta_j, \eta_{j+1}\right]$. The NURBS basis functions $R_{i,j}^{p,q}(\xi, \eta)$ are defined over the parametric space. Each element in the physical space depicts the image of the corresponding element in the parametric space. The parametric space for a particular knot vector is shown in Fig. 1c.

The parent element describes a constant space $[-1, 1] \times [-1, 1]$ where the integration is performed for each element. The parameters ξ and η of the parametric space are mapped onto the $\hat{\xi}$ and $\hat{\eta}$ in the parent element so that the Gauss quadrature can be used for integration. A typical parent element is shown in Fig. 1d.

In IGA, two transformations are performed from physical space to the parent element for the generation of Gauss points. First, the physical space is transformed into parametric space, which is constructed from the knot spans. The parametric space is then transformed into parent element.

3 Extended Isogeometric Analysis

The NURBS basis functions are suitable for the problems requiring smooth solutions due to their higher order continuity. However, for modeling the problems involving defects (crack, holes and inclusions), IGA approximation enriched using partition of unity (PU) is known as extended isogeometric analysis (XIGA). PU enrichment is quite effective and efficient method for analyzing static and moving discontinuities in the structures, whereas IGA is quite accurate and efficient in analyzing complex geometries [15]; hence, the merits of both XFEM (PU enrichment) and IGA are exploited in XIGA for the accurate modeling of the components/structures. The isogeometric formulation for the analysis of structure/component is explained in [5, 6].

3.1 XIGA Approximation for Cracks

At a particular control point corresponding to ξ_i in the parametric space, the displacement approximation for cracks can be written in a generalized form as,

$$\mathbf{u}^{\mathrm{h}}(\xi) = \sum_{i=1}^{n_{en}} R_i(\xi)\mathbf{u}_i + \left\{ \sum_{j=1}^{n_{cf}} R_j(\xi)\left[H(\xi) - H(\xi_i)\right]\mathbf{a}_j \right\} B_H + \left[\sum_{k=1}^{n_{ct}} R_k(\xi) \left\{ \sum_{\alpha=1}^{4} \left[\beta_\alpha(\xi) - \beta_\alpha(\xi_i)\right] \mathbf{b}_k^\alpha \right\} \right] B_T \tag{4}$$

where, $\mathbf{u}_i$ = parameters associated with a particular control point,n_{en} = number of basis functions per element defined as $n_{en} = (p+1) \times (q+1)$, n_{cf} = number of basis functions whose support is completely intersected by the crack face, n_{ct} = number of basis functions whose support is partially intersected by the crack tip, $\mathbf{a}_j$ = additional control point enriched degrees of freedom associated with Heaviside function H(ξ), $\mathbf{b}_k^\alpha$ = additional control point enriched degrees of freedom associated with asymptotic crack tip enrichment functions, β_α, B_H = blending function for the Heaviside function, B_T = blending function for the crack tip, $H(\xi)$ = Heaviside function, β_α = crack-tip enrichment functions.

At a particular Gauss point in an element, if the parametric coordinate ξ corresponding to that particular Gauss point is above the crack face, then $H(\xi)$ is equal to $+1$ and if ξ lies below the crack face, then $H(\xi)$ is -1. Similarly, discontinuous Heaviside function $H(\xi_i)$ takes value $+1$ and -1 if a control point lies above and below the crack face respectively.

3.2 XIGA Formulation

As mentioned above, Eq. (4) defines the enriched displacement approximation at a particular point in XIGA. The first term on right-hand side of Eq. (4) represents the standard finite element approximation, while the remaining enrichment terms are used to model crack. Using the approximation defined in Eq. (4), the elemental matrices **k** and **f** are defined as,

$$\mathbf{k}_{ij}^{e} = \begin{bmatrix} \mathbf{k}_{ij}^{uu} & \mathbf{k}_{ij}^{ua} & \mathbf{k}_{ij}^{ub} \\ \mathbf{k}_{ij}^{au} & \mathbf{k}_{ij}^{aa} & \mathbf{k}_{ij}^{ab} \\ \mathbf{k}_{ij}^{bu} & \mathbf{k}_{ij}^{ba} & \mathbf{k}_{ij}^{bb} \end{bmatrix} \quad \text{and} \quad \mathbf{f}_{i}^{e} = \left\{ \mathbf{f}_{i}^{u} \quad \mathbf{f}_{i}^{a} \quad \mathbf{f}_{i}^{b1} \quad \mathbf{f}_{i}^{b2} \quad \mathbf{f}_{i}^{b3} \quad \mathbf{f}_{i}^{b4} \right\}^{T} \tag{5}$$

where,

$$\mathbf{k}_{ij}^{rs} = \int_{\Omega^e} (\mathbf{B}_i^r)^T \mathbf{C} \mathbf{B}_j^s d\Omega, \mathbf{f}_i^u = \int_{\Omega^e} \mathbf{R}_i^T \mathbf{b} d\Omega + \int_{\Gamma_t} \mathbf{R}_i^T \hat{\mathbf{t}} d\Gamma \tag{6}$$

$$\mathbf{f}_i^a = \int_{\Omega^e} \mathbf{R}_i^T H \mathbf{b} d\Omega + \int_{\Gamma_t} \mathbf{R}_i^T H \, \hat{\mathbf{t}} d\Gamma \quad , \quad \mathbf{f}_i^{b\alpha} = \int_{\Omega^e} \mathbf{R}_i^T \beta_\alpha \mathbf{b} d\Omega + \int_{\Gamma_t} \mathbf{R}_i^T \beta_\alpha \hat{\mathbf{t}} d\Gamma \tag{7}$$

where, R_i represents the NURBS basis function, and $\mathbf{B}_i^u$, $\mathbf{B}_i^a$, $\mathbf{B}_i^b$, $\mathbf{B}_i^{b\alpha}$, $\mathbf{B}_i^c$, and $\mathbf{B}_i^d$ are the matrices of derivatives of NURBS basis function.

$$\mathbf{B}_i^u = \begin{bmatrix} R_{i,X_1} & 0 \\ 0 & R_{i,X_2} \\ R_{i,X_2} & R_{i,X_1} \end{bmatrix} \quad \mathbf{B}_i^a = \begin{bmatrix} (R_i)_{,X_1} H & 0 \\ 0 & (R_i)_{,X_2} H \\ (R_i)_{,X_2} H & (R_i)_{,X_1} H \end{bmatrix} \quad \mathbf{B}_i^b = \begin{bmatrix} \mathbf{B}_i^{b1} & \mathbf{B}_i^{b2} & \mathbf{B}_i^{b3} & \mathbf{B}_i^{b4} \end{bmatrix}$$

For the details on the selection of enriched control points, integration in discontinuous elements, computation of stress intensity factors, fatigue crack growth criterion, and Paris law, please refer [6].

In the present work, maximum hoop (circumferential) stress criterion [8] is used for crack growth direction.

$$\theta_c = 2 \tan^{-1} \left(\frac{K_{\mathrm{I}} - \sqrt{K_{\mathrm{I}}^2 + 8K_{\mathrm{II}}^2}}{4K_{\mathrm{II}}} \right) \tag{8}$$

4 Fatigue Crack Growth in Bi-material

The fatigue crack growth analysis of an interfacial edge crack or center crack in bi-material has been performed by both XIGA and XFEM under plane stress condition. The domain containing an edge or center crack is subjected to tensile load of $\sigma_{min} = 0$ MPa and $\sigma_{max} = 70$ MPa at the top edge, whereas the bottom edge is constrained in y-direction. Each material comprising the bi-material is assumed to be homogenous and isotropic. The upper half of the domain consists of low stiffness material (material-1 or $m1$), whereas the lower half consists of a high stiffness material (material-2 or $m2$). The material properties of Material-1 ($m1$) and Material-2 ($m2$) are given in Table 1. The numerical simulations are performed using a control net of 30 × 60 in case of XIGA and a mesh size of 50 × 100 in case of XFEM.

4.1 Edge Crack Growth

A bi-material rectangular domain of dimension 50 mm × 100 mm with an interfacial edge crack of initial length, $a = 5$ mm, is taken for the fatigue crack growth analysis as shown in Fig. 2.

The normal stress (σ_{yy}) contour plot for a crack length of 5.0 mm is shown in Fig. 3. The variation of equivalent SIF with crack extension is shown in Fig. 4. The crack growth direction is obtained from Eq. (8), and then the crack is extended by 2 mm in that direction. The problem is again solved for new crack length, and SIFs values and crack growth direction are obtained again. This process continues till the final failure i.e., simulation stops when $K_{Ieq} \geq \min\left\{(K_{IC})_{m1}, (K_{IC})_{m2}\right\}$. The fatigue life of bi-material cracked domain evaluated by XIGA using Paris law is presented in Fig. 5. The failure life and crack length (at the time of failure) are found to be 3.1×10^6 cycles and 14 mm, respectively.

Table 1 Material properties of aluminum alloy ($m1$) [1] and steel ($m2$) [11]

Property	Aluminum (m1)	Steel (m2)
Elastic modulus (GPa)	70	200
Poisson's ratio	0.33	0.30
Fracture toughness ($\text{MPa}\sqrt{\text{mm}}$)	917.06	1897.36
Paris law constant m/cycle $\left(\text{MPa}\sqrt{\text{m}}\right)^{-M}$	1×10^{-12}	2.087136×10^{-12}
Paris law exponent	3	3

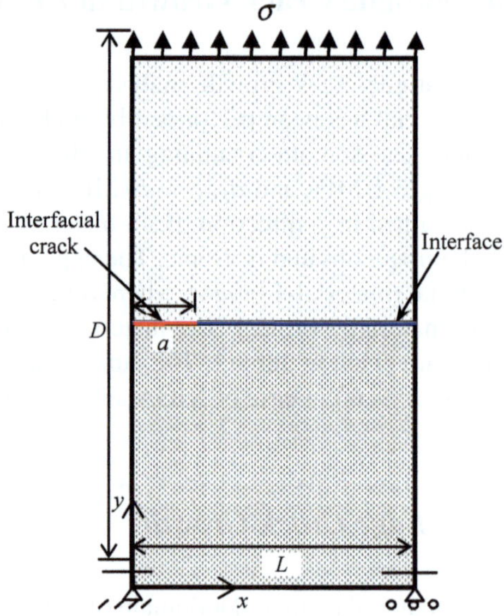

Fig. 2 Interfacial edge cracked bi-material body with loading and boundary conditions

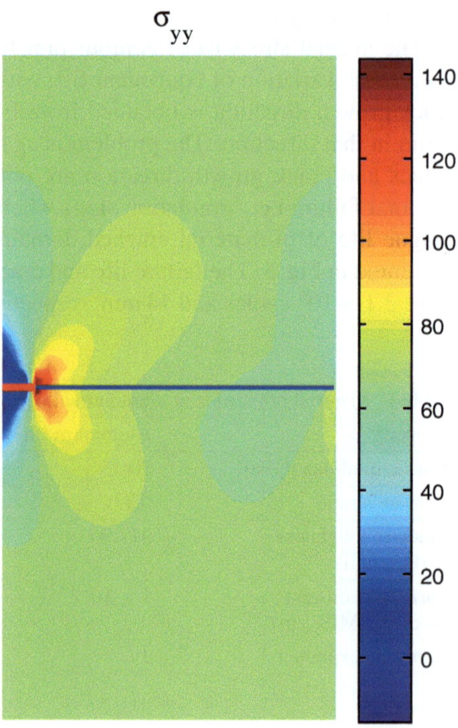

Fig. 3 Contour plot of σ_{yy} (MPa) for an interfacial edge crack in bi-material

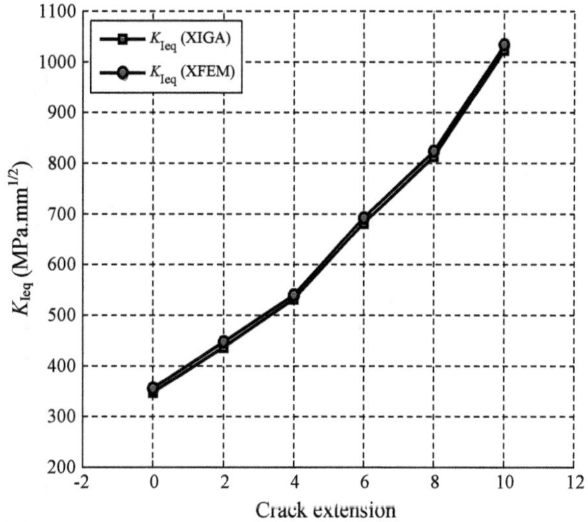

Fig. 4 K_{Ieq} variation with crack extension for an interfacial edge crack in bi-material

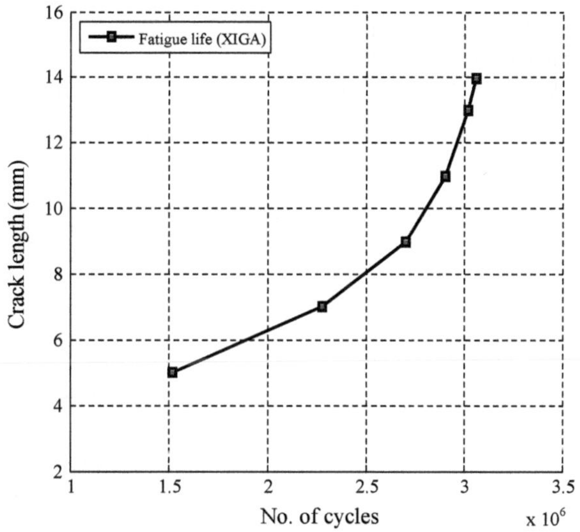

Fig. 5 Fatigue life variation with crack length for an interfacial edge crack in bi-material

4.2 Center Crack Growth

A bi-material rectangular domain of size 50 mm × 100 mm with an interfacial center crack of length, $2a = 10$ mm, is taken for analysis as shown in Fig. 6. The contour plot of normal stress (σ_{yy}) for a crack length of 10 mm is shown in Fig. 7. The variation of equivalent SIF (K_{Ieq}) with crack length is shown in Fig. 8. The crack

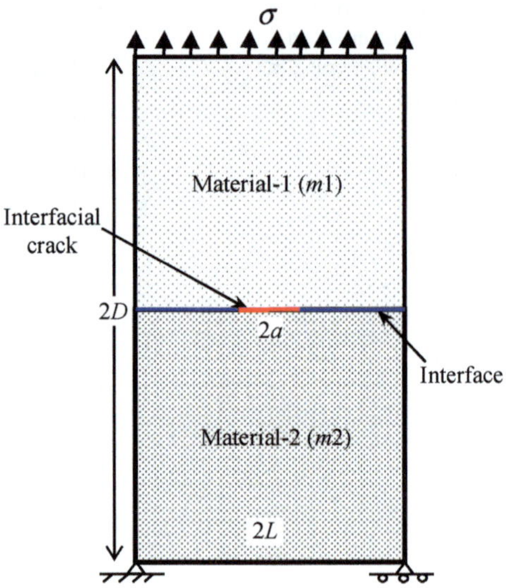

Fig. 6 Interfacial center cracked bi-material body with loading and boundary conditions

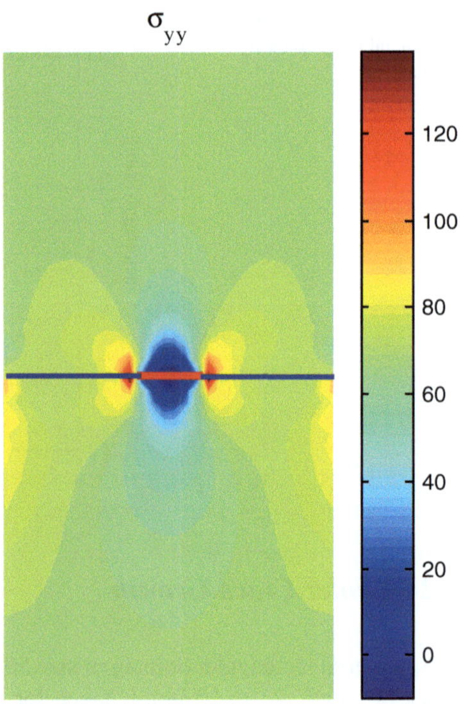

Fig. 7 Contour plot of σ_{yy} (MPa) for an interfacial center crack in bi-material

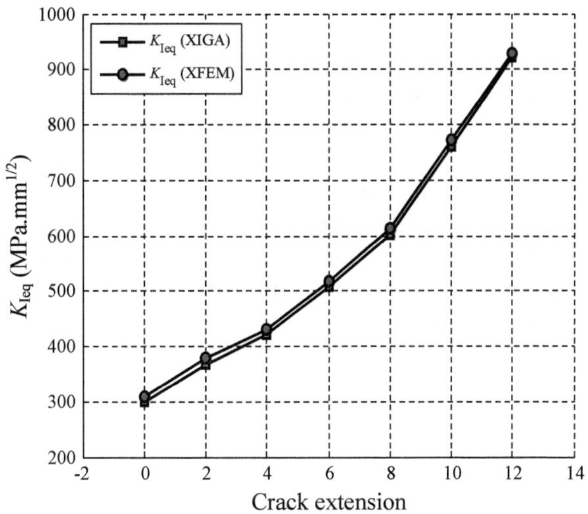

Fig. 8 K_{Ieq} variation with crack extension for an interfacial center crack in bi-material

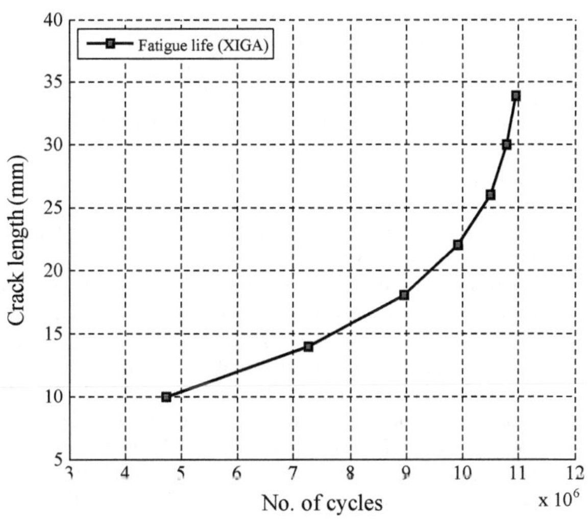

Fig. 9 Fatigue life variation with crack length for an interfacial center crack in bi-material

growth direction is obtained from Eq. (8), and then crack is extended by 2 mm in that direction. The problem is again solved for new crack length, and SIFs values and crack growth direction are obtained again. This process continues till the final failure i.e., simulation stops when $K_{\text{Ieq}} \geq \min \left\{ (K_{\text{IC}})_{m1}, (K_{\text{IC}})_{m2} \right\}$.

The fatigue life of bi-material cracked body evaluated by XIGA using Paris law is shown in Fig. 9. The failure life and corresponding crack length at the time of failure are found to be 1.1×10^7 cycles and 33.9 mm.

5 Conclusions

In this work, the IGA is extended to solve interfacial fracture mechanics problems. The fatigue crack growth analysis of edge or center crack in bi-material domain is performed under mode-I loading. The mode-I SIF values obtained by XIGA (using a control net of 30×60) are compared with the XFEM using a mesh size of 50×100). In bi-material, the crack kinks out from the interface and travels in material with lower fracture toughness along the interface. These simulations show that the results obtained by XIGA using coarse control net are in good agreement with those obtained by XFEM using fine mesh.

References

1. Arola, D., Huang, M.P., Sultan, M.B.: The failure of amalgam dental restorations due to cyclic fatigue crack growth. J. Mater. Sci.: Mater. Med. **10**, 319–327 (1999)
2. Belytschko, T., Black, T.: Elastic crack growth in finite elements with minimal remeshing. Int. J. Numer. Methods Eng. **45**, 601–620 (1999)
3. Belytschko, T., Gu, L., Lu, Y.Y.: Fracture and crack growth by element free Galerkin methods. Model. Simul. Mater. Sci. Eng. **2**, 519–534 (1994)
4. Bhardwaj, G., Singh, I.V., Mishra, B.K.: Stochastic fatigue crack growth simulations of interfacial crack in bi-layered FGM's using XIGA. Comput. Methods Appl. Mech. Eng. **284**, 186–229 (2015)
5. Bhardwaj, G., Singh, I.V., Mishra, B.K.: Numerical simulation of functionally graded cracked plates using NURBS based XIGA under different loads and boundary conditions. Compos. Struct. **126**, 347–359 (2015)
6. Bhardwaj, G., Singh, S.K., Singh, I.V., Mishra, B.K., Rabczuk, T.: Fatigue crack growth analysis of an interfacial crack in heterogeneous materials using homogenized XIGA. Theor. Appl. Fract. Mech. **85**, 294–319 (2016)
7. Cottrell, J.A., Hughes, T.J.R., Reali, A.: Studies of refinement and continuity in isogeometric structural analysis. Comput. Methods Appl. Mech. Eng. **196**, 4160–4183 (2007)
8. Erdogan, F., Sih, G.C.: On the crack extension in plates under plane loading and transverse shear. ASME J. Fluids Eng. **85**, 519–525 (1963)
9. Hughes, T.J.R., Cottrell, J.A., Bazilevs, Y.: Isogeometric analysis: CAD, finite elements, NURBS, exact geometry and mesh refinement. Comput. Methods Appl. Mech. Eng. **194**, 4135–4195 (2005)
10. Liu, W.K., Jun, S., Zhang, Y.F.: Reproducing kernel particle methods. Int. J. Numer. Methods Eng. **20**, 1081–1106 (1995)
11. Pathak, H., Singh, A., Singh, I.V.: Fatigue crack growth simulations of bi-material interfacial cracks under thermo elastic loading by extended finite element method. Eur. J. Comput. Mech. **22**, 79–104 (2013)
12. Rabczuk, T., Belytschko, T.: Cracking particles: a simplified meshfree method for arbitrary evolving cracks. Int. J. Numer. Methods Eng. **61**, 2316–2343 (2004)
13. Singh, I.V., Bhardwaj, G., Mishra, B.K.: A new criterion for modeling multiple discontinuities passing through an element using XIGA. J. Mech. Sci. Technol. **29**, 1131–1143 (2015)
14. Yan, A.M., Nguyen-Dang, H.: Multiple-cracked fatigue crack growth by BEM. Comput. Mech. **16**, 273–280 (1995)
15. Wall, W.A., Frenzel, M.A., Cyron, C.: Isogeometric structural shape optimization. Comput. Methods Appl. Mech. Eng. **197**, 2976–2988 (2008)

A Comparison of Delamination Models: Modeling, Properties, and Applications

Marita Thomas

Abstract This contribution presents recent results in the modeling and the analysis of delamination problems. It addresses adhesive contact, brittle, and cohesive zone models both in a quasistatic and a viscous, dynamic setting for the bulk part. Also different evolution laws for the delaminating surface are discussed.

1 Continuum-Mechanical Modeling Approaches to Fracture

The creation and growth of cracks in a solid body corresponds to the formation of new (interior) surfaces. From a continuum-mechanical modeling point-of-view, the field variables describing the state of the body may display discontinuities along these crack surfaces. To mathematically model an elastic body with an evolving crack one may formulate the system of elasticity on a moving domain characterized by the propagation of the crack surface, which, in turn, has to be described by a suitable evolution law. Over the decades, in literature, many different criteria have been proposed to determine the unknown direction for the extension of an existing crack, famous among them are the maximum energy release rate criterion and the maximum stress criterion, see e.g. [40, 58] for a discussion. However, to handle the formation of discontinuities in general geometries in a mathematically rigorous way requires the formulation of the problem in adequate function spaces, such as $GSBV$-spaces ($GSBD$-spaces) of functions of generalized special bounded variations [4] (deformations [17]), as used e.g. in the Francort–Marigo model for quasistatic brittle fracture [23]. Instead, many models make regularizing assumptions on the discontinuity set. For this, often an additional internal variable is introduced in the spirit of generalized standard materials [29], with the purpose to describe the changes in the elastic behavior of the material caused by the evolving inelastic process. This is the basis for the phase field approach to fracture, where the phase field variable

M. Thomas (✉)
Weierstrass Institute for Applied Analysis and Stochastics,
Mohrenstr. 39, 10117 Berlin, Germany
e-mail: thomas@wias-berlin.de

P. van Meurs et al. (eds.), *Mathematical Analysis of Continuum Mechanics and Industrial Applications II*, Mathematics for Industry 30,
https://doi.org/10.1007/978-981-10-6283-4_3

regularizes the model by replacing the lower dimensional crack *surface* by a damaged *volume* of controllable width, where the displacement field displays no discontinuities as associated with an elastic solid. Starting from the Ambrosio–Tortorelli phase field model, which was shown in [27] to approximate the Francort–Marigo model in the sense of Γ-convergence, phase field models for facture have received much attention both in mathematical analysis and numerical simulation, see e.g. [1, 12, 15, 16, 24, 28, 55]. Since they allow it to capture complicated geometric situations, such as crack initiation and branching, they have become a well-established method in a wide range of engineering applications, cf. e.g. [3, 30, 33–35, 54, 59].

In case of *compounds* of elastic solids, with applications ranging from laminates to layers of rocks and soil, fracture often occurs in terms of *delamination*, i.e. cracks form and propagate along the interface between two material layers. In such a situation of delamination, one may resort to models, where crack initiation and propagation are confined to a prescribed interface $\Gamma_C \subset \mathbb{R}^{d-1}$ between two parts Ω_+, Ω_- of an elastic solid $\Omega \subset \mathbb{R}^d$. The domain of the internal variable, which then describes the state of the bonding along the interface, is consequently confined to Γ_C.

This setup will be considered in the present contribution, which is devoted to the discussion of recent results in modeling and mathematical analysis of delamination processes. Section 2 explains the mathematical modeling at the example of *adhesive contact* and gives an overview on existence results developed in this field. In Sect. 3, the modeling and analytical challenges of adhesive contact are compared with those of *brittle delamination* and *cohesive zone delamination*.

2 Modeling of Delamination Processes via Energy and Dissipation Functionals at the Case of Adhesive Contact

Thermodynamically consistent modeling of adhesive contact via internal variables goes back to [25], cf. also [26]. Therein, the internal delamination variable $z : [0, T] \times \Gamma_C \to [0, 1]$ describes the state of the bonding along the interface Γ_C between the two parts $\Omega_+, \Omega_- \subset \mathbb{R}^d$ of the body, monitored during a finite time interval $[0, T]$. In what follows, $z(t, x) = 1$ will indicate that the bonding at the point $x \in \Gamma_C$ at time $t \in [0, T]$ is fully intact, whereas $z(t, x) = 0$ means that the bonding is completely broken, and $z(t, x) \in (0, 1)$ stands for an intermediate state of degradation. In this way, the set $C(t) := \{x \in \Gamma_C,\ z(t, x) = 0\}$ defines the crack at time $t \in [0, T]$.

In the isothermal setting at small strains the state of a viscoelastic body and an adhesive is thus described by the pair of state variables (u, z) consisting of the displacement field $u : [0, T] \times \Omega\backslash\Gamma_C \to \mathbb{R}^d$ and the delamination variable $z : [0, T] \times \Gamma_C \to [0, 1]$. Their evolution is governed in terms of a tuple $(\mathbf{V}, \mathbf{W}, \mathbf{Z}, \mathscr{V}, \mathscr{K}, \mathscr{R}, \mathscr{E})$ given by suitable state spaces $\mathbf{V}, \mathbf{W}, \mathbf{Z}$, a stored energy $\mathscr{E}$, the kinetic energy $\mathscr{K}$, a dissipation potential $\mathscr{V}$ accounting for viscosity in the bulk, and a dissipation potential $\mathscr{R}$ accounting for the energy dissipated due to the delamination process.

The stored energy functional $\mathscr{E} := \mathscr{E}_{\text{bulk}} + \mathscr{E}_{\text{adh}}$ is composed of a bulk term defined on $\Omega\backslash\Gamma_{\text{C}} = \Omega_+ \cup \Omega_-$ and a surface term defined on Γ_{C}. Characteristically for adhesive contact models, which additionally also account for non-penetration of the material along Γ_{C}, the latter is given as follows:

$$\mathscr{E}_{\text{adh}} : \mathbf{V} \times \mathbf{Z} \to [0, \infty], \;\; \mathscr{E}_{\text{adh}}(u, z) := \int_{\Gamma_{\text{C}}} \Big(\tfrac{k}{2} z |[\![u]\!]|^2 + I_{[0,1]}(z) + I_{[0,\infty)}\Big([\![u]\!]_{\mathbf{n}}\Big)\Big) \, \mathrm{d}\mathscr{H}^{d-1} . \tag{1}$$

Here, $k > 0$ is fixed and $\mathscr{H}^{d-1}$ denotes the $(d-1)$-dimensional Hausdorff-measure. Furthermore, $[\![u]\!] := u_+ - u_-$ denotes the jump of the displacements across Γ_{C} calculated from the traces u_+, u_- along Γ_{C} of the functions $u|_{\Omega_+}, u|_{\Omega_-}$. In a similar manner, $[\![u]\!]_{\mathbf{n}} := [\![u]\!] \cdot \mathbf{n}$ is the jump of the displacements in normal direction to Γ_{C} with the normal defined as the outward normal to the subdomain Ω_+. The constraint $z \in [0, 1]$ is ensured by the indicator function $I_{[0,1]} : \mathbb{R} \to \{0, \infty\}$ of the interval $[0, 1]$, i.e., $I_{[0,1]}(z) = 0$ if $z \in [0, 1]$ and $I_{[0,1]}(z) = \infty$ otherwise. Finally, due to the term $I_{[0,\infty)}([\![u]\!]_{\mathbf{n}})$, a finite surface energy $\mathscr{E}_{\text{adh}}(u, z)$ corresponds to $[\![u]\!]_{\mathbf{n}} \geq 0$ a.e. on Γ_{C}, and thus forbids the penetration of matter along Γ_{C}.

For a linear elastic material the bulk energy may be considered as

$$\mathscr{E}_{\text{bulk}} : [0, T] \times \mathbf{V} \to \mathbb{R}, \;\; \mathscr{E}_{\text{bulk}}(t, u) := \int_{\Omega\backslash\Gamma_{\text{C}}} \big(\tfrac{1}{2}\mathbb{C}e(u) : e(u) - f(t) \cdot u\big) \, \mathrm{d}x \tag{2}$$

with $\mathbb{C} \in \mathbb{R}^{d\times d\times d\times d}$ a symmetric, positively definite fourth-order tensor and $e(u) := \frac{1}{2}(\nabla u + \nabla u^\top)$ the small strain tensor. The evolution process is driven here by a time-dependent volume force f. The kinetic energy $\mathscr{K}$ is given by

$$\mathscr{K} : \mathbf{W} \to [0, \infty), \;\; \mathscr{K}(v) := \frac{1}{2}\|v\|^2_{\mathbf{W}} \text{ with } \|v\|^2_{\mathbf{W}} := \int_\Omega \rho |v|^2 \mathrm{d}x \tag{3}$$

for some given function $\rho \in L^\infty(\Omega)$ and a constant $\rho_* > 0$ such that $\rho > \rho_*$ a.e. in Ω. In this way, the space $\mathbf{W}$ is equivalent to $L^2(\Omega; \mathbb{R}^d)$.

The dissipation potential $\mathscr{V} : \mathbf{V} \to [0, \infty)$ accounts for viscosity of the bulk material and is here for simplicity assumed to be quadratic:

$$\mathscr{V} : \mathbf{V} \to [0, \infty), \;\; \mathscr{V}(v) := \int_{\Omega\backslash\Gamma_{\text{C}}} \tfrac{1}{2}\mathbb{D}e(v) : e(v) \, \mathrm{d}x \, , \tag{4}$$

again, with $\mathbb{D} \in \mathbb{R}^{d\times d\times d\times d}$ a symmetric, positively definite fourth-order tensor. Assuming here for simplicity homogeneous Dirichlet conditions along the outer boundary $\partial\Omega$, in view of (2) and (4), it is $\mathbf{V} := \{v \in H^1(\Omega\backslash\Gamma_{\text{C}}),\, v = 0 \text{ a.e. on } \partial\Omega\}$.

The dissipation potential $\mathscr{R}$ controls how much energy is spent on the progression of delamination. In many materials, such as metals or rocks, fracture is a unidirectional progress, in the sense that cracks can only grow but not heal; this feature may be encoded in the dissipation potential by considering $\mathscr{R} : \mathbf{Z} \to [0, \infty]$,

$$\mathscr{R}(v) := \widetilde{\mathscr{R}}(v) + \alpha \int_{\Gamma_C} I_{(-\infty,0]}(v)\,\mathrm{d}\mathscr{H}^{d-1} \text{ with } \widetilde{\mathscr{R}}(v) := \int_{\Gamma_C} \widetilde{R}(v)\,\mathrm{d}\mathscr{H}^{d-1}, \quad (5)$$

where $\widetilde{R} : \mathbb{R} \to [0, \infty)$ is assumed to be convex with $\widetilde{R}(0) = 0$ and where $I_{(-\infty,0]} : \mathbb{R} \to \{0, \infty\}$ denotes the indicator function of $(-\infty, 0]$ and α is a non-negative constant. Taking $v = \dot{z}$, in view on the modeling assumptions on z, this captures that a growth of delamination corresponds to a decrease of z in time (from the value 1 at most down to 0), so that, formally, its time-derivative always should be non-increasing.

Formally, the evolution of the state variables (u, z) is then captured by the system

$$\rho\ddot{u} - \operatorname{div}\big(\mathbb{C}e(u) + \mathbb{D}e(\dot{u})\big) = f \text{ in } [0, T] \times \Omega\backslash\Gamma_C\,, \quad (6a)$$

$$u = 0 \text{ on } [0, T] \times \partial\Omega\,, \quad (6b)$$

$$\big[\!\big[\mathbb{C}e(u) + \mathbb{D}e(\dot{u})\big]\!\big] \cdot \mathbf{n} = 0 \text{ on } [0, T] \times \Gamma_C\,, \quad (6c)$$

$$\big(\mathbb{C}e(u) + \mathbb{D}e(\dot{u})\big) \cdot \mathbf{n} + kz[\![u]\!] + \partial_{[\![u]\!]} I_{[0,\infty)}\big([\![u]\!]_{\mathbf{n}}\big) \ni 0 \text{ on } [0, T] \times \Gamma_C\,, \quad (6d)$$

$$\partial_{\dot{z}}\big(\widetilde{R}(\dot{z}) + \alpha I_{(-\infty,0]}(\dot{z})\big) + \tfrac{k}{2}\big|[\![u]\!]\big|^2 + \partial_z I_{[0,1]}(z) \ni 0 \text{ on } [0, T] \times \Gamma_C\,, \quad (6e)$$

$$\nabla z \cdot \mathbf{n} = 0 \text{ on } [0, T] \times \Gamma_C\,, \quad (6f)$$

and supplemented by initial conditions for (u, z).

The indicator terms I_K of the convex intervals $K \in \{[0, 1], (-\infty, 0], [0, \infty)\}$ appearing in the functionals (1) and (5) are highly non-smooth, which is why (6d) and (6e) are given in terms of subdifferential inclusions and display the multivalued subdifferentials ∂I_K of the corresponding indicator function. Here, the subdifferential $\partial g(v)$ of a convex function $g : \mathbb{R} \to [0, \infty]$ at v is given by the set of all subderivatives v^*, i.e., $\partial g(v) := \{v^* \in \mathbb{R},\ g(w) - g(v) \geq v^*(w - v) \text{ for all } w \in \mathbb{R}\}$, which is the set of "slopes" defining tangents to g in the point v. In particular, if a function is classically differentiable in a point v, then the subdifferential is single-valued in v and coincides with the derivative in v. For example, the subdifferential of $I_{[0,1]}$ is given by

$$\partial I_{[0,1]}(z) = \begin{cases} \emptyset & \text{if } z \in (-\infty, 0) \cup (1, \infty), \\ (-\infty, 0] & \text{if } z = 0, \\ \{0\} & \text{if } z \in (0, 1), \\ [0, \infty) & \text{if } z = 1 \end{cases}, \quad (7)$$

i.e., the subdifferential is multivalued and, in general, given by an unbounded set. Exactly this bears the main difficulty in the analysis of system (6) and suitable weak notions of solution in suitable functions spaces as well as refined analytical tools have to be established in order to deduce suitable bounds on each of the occuring terms. To be more precise, a first major difficulty comes with the momentum balance (6a) and the corresponding transmission condition (6d) along Γ_C. A weak formulation combines the acceleration term with elements of the unbounded subdifferential of the non-penetration constraint and, in general, bounds needed for the analysis can be deduced only on the sum of the two but not on each of the two terms separately. Many works dealing with adhesive contact and involving the non-penetration constraint thus resort to a physical setting with external loadings varying sufficiently slowly so that acceleration terms can be neglected, cf. e.g. [7–10], or replace the non-penetration constraint by less restrictive conditions, cf. e.g. [43], where tangential slip is admissible. Without the acceleration term it is possible, in particular, to give a weak formulation of (6a)–(6d) in $L^2(0,T;\mathbf{V})$ and to obtain suitable bounds on all terms in this setting, however, as indicated, at the price of neglecting the acceleration term. Only recently, in a series of papers [11, 51, 53], it was indeed possible to handle dynamics in combination with the non-penetration constraint using a weak formulation in $H^1(0,T;\mathbf{V})$, instead. This opens the door for a number of additional, refined estimates, also using test functions, which incapacitate the element of the unbounded subdifferential, so that a separate estimate on the acceleration term becomes accessible. The approach of resorting to smaller spaces of admissible test functions, with properties tightly tailored to handle unbounded terms arising from non-smooth constraints seems to be the key to deal with rate-dependence and dynamics in combination with non-smooth constraints.

Let us now discuss the difficulties related to the flow rule (6e). In literature, two major cases can be distinguished for the density $\widetilde{R}$ involved in the dissipation potential $\mathscr{R}$ from (5): The case, where $\widetilde{R}$ contains a quadratic term and the case where $\widetilde{R}$ is positively 1-homogeneous. As a first step, let us assume for the moment that $\widetilde{R}$ is quadratic, i.e., $\widetilde{R}(v) := \frac{1}{2}|v|^2$. Then, we may choose the space $\mathbf{Z} := L^2(\Gamma_{\mathrm{C}})$ and the flow rule (6e) can be rewritten as follows:

$$\dot{z} + \tfrac{k}{2}\big|[\![u]\!]\big|^2 + \xi \le 0\,, \tag{8a}$$
$$\dot{z}(\dot{z} + \tfrac{k}{2}\big|[\![u]\!]\big|^2 + \xi) = 0\,, \tag{8b}$$
$$\dot{z} \le 0\,, \tag{8c}$$
$$\xi \in \partial_z I_{[0,1]}(z)\,. \tag{8d}$$

For $z = 1$, by (7), it is $\xi \in [0,\infty)$. Thus, as long as $\dot{z} = 0$, by (8), it has to hold $\frac{k}{2}\big|[\![u]\!]\big|^2 = \xi = 0$. Delamination sets in, i.e., $\dot{z} < 0$, if and only if $\xi = -(\dot{z} + \frac{k}{2}\big|[\![u]\!]\big|^2)$. As soon as a weakening of the bonds has taken place, so that $z \in (0,1)$, according to (7), it is $\xi = 0$ and then, again by (8), delamination remains static, i.e., $\dot{z} = 0$ as long as $\frac{k}{2}\big|[\![u]\!]\big|^2 = 0$ as well. In turn, delamination will progress any further, i.e., $\dot{z} < 0$, if and only if $\dot{z} = -\frac{k}{2}\big|[\![u]\!]\big|^2$. Once the final state of rupture $z = 0$ is reached, we see that also $\dot{z} = 0$ is feasible, because $\xi \le -\frac{k}{2}\big|[\![u]\!]\big|^2$ is admissible for any value of $\frac{k}{2}\big|[\![u]\!]\big|^2$ thanks to $\xi \in (-\infty, 0]$ by (7). In other words, the temporal evolution (considered pointwise in Γ_{C}) can be interpreted to stop once the value $z = 0$ is reached. An evolution of adhesive contact of viscous kind has been treated e.g. in

[52, 53] and also in [7, 9, 10], the latter works neglect unidirectionality, i.e., $\alpha = 0$ in (5) and (6e).

For comparison, let us now consider a positively 1-homogeneous density $\widetilde{R}$, i.e. $\widetilde{R}(v) = a|v|$ with a constant $a > 0$ and hence we now set $\mathbf{Z} := L^1(\Gamma_\mathrm{C})$. Clearly, this $\widetilde{R}$ is convex but not classically differentiable in $v = 0$ and its subdifferential in $v = 0$ is given by $\partial R(0) = [-a, a]$. In combination with the unidirectionality constraint with $\alpha = 1$, we have that $R = a|v| + I_{(-\infty,0]}(v)$ in (5). Accordingly, (6e) amounts to the following conditions

$$\text{(8c) and (8d) combined with } \tfrac{k}{2}\big|[\![u]\!]\big|^2 + \xi \leq a\,, \tag{9a}$$

$$\text{and} \quad \dot{z}\big(\tfrac{k}{2}\big|[\![u]\!]\big|^2 + \xi - a\big) = 0\,. \tag{9b}$$

For $z = 1$, by (7), $\xi \in [0, \infty)$ is possible. Condition (9a) then imposes a threshold, below which $\dot{z} = 0$ has to hold in order to fulfill (9b). When the threshold is reached, i.e., $\frac{k}{2}\big|[\![u]\!]\big|^2 + \xi = a$, delamination may set in with $\dot{z} \leq 0$. Once the bonds are weakened, $z \in (0, 1)$, it is $\xi = 0$. Then delamination may propagate as soon as the displacement jump reaches the threshold $\frac{k}{2}\big|[\![u]\!]\big|^2 = a$. In the final state of rupture $z = 0$, (7) yields $\xi \in (-\infty, 0]$, so that $\frac{k}{2}\big|[\![u]\!]\big|^2 + \xi < a$ can be satisfied in (9) for any value of $\frac{k}{2}\big|[\![u]\!]\big|^2$. Thus, $\dot{z} = 0$ must be assumed to fulfill (9b), and hence the temporal evolution (pointwise in Γ_C) can be interpreted to stop once the value $z = 0$ is reached. Comparing the evolution conditions (9) induced by a positively 1-homogeneous dissipation potential with (8) given by a quadratic potential, one observes the following: In (9) the attainment of the threshold is independent of the rate $\dot{z}$ and also (9b) only distinguishes between $\dot{z} = 0$ and $\dot{z} \neq 0$, but the particular values of the rate $\dot{z}$ do not play any role. In contrast, in the corresponding equations of (8) the rate $\dot{z}$ takes an explicit influence. In this way, a *quadratic* dissipation potential describes a *rate-dependent* evolution, whereas a *positively* 1-*homogeneous* dissipation potential governs a *rate-independent* evolution of the internal variable.

From a mathematical point-of-view rate-independence entails that formulation (6e), resp. (9), cannot provide any bounds on the rate $\dot{z}$ in Lebesgue–Bochner spaces. Indeed, one observes that solutions are measure-valued in time, so that, in fact, solutions may jump in time, see e.g. [38, 39, 56] for results on the temporal regularity of rate-independent processes with quasistatic evolution of u and also [47] for the coupled rate-dependent/rate-independent situation as in (6). Due to this lack of regularity in time, in general, formulations (6e) and (9) are not well-defined and, instead, one has to resort to an alternative notion of solution, which does not explicitly involve the time derivative. To motivate this alternative notion of solution, observe that (9b) states a power balance, while (9a) defines a local stability criterion of the form: $\langle \mathrm{D}_z\mathscr{E}(t, u(t), z(t)), v\rangle + \mathscr{R}(v) \geq 0$ for all suitable test functions v. Taking $v = \tilde{z} - z(t)$ and exploiting the convexity wrt. z of the energy functional $\mathscr{E} = \mathscr{E}_\mathrm{bulk} + \mathscr{E}_\mathrm{adh}$ with $\mathscr{E}_\mathrm{adh}$, one arrives at the *semistability* condition for all $t \in [0, T]$:

$$\text{For all test functions } \tilde{z}: \quad \mathcal{E}(t, u(t), z(t)) \leq \mathcal{E}(t, u(t), \tilde{z}) + \mathcal{R}(\tilde{z} - z(t)) . \tag{10}$$

Here, the notion of *semistability* indicates that (10) is a stability condition only for z, whereas u is determined by a weak formulation of the momentum balance (6a). In addition, formally, we may test the weak formulation of (6a) by $\dot{u}$. Note then, that the sum of this and the power balance (9b) state a power balance for the full system. Then, formally, we may integrate by parts in time to obtain an energy-dissipation balance for the system:

$$\mathcal{E}(t, u(t), z(t)) + \int_0^t \int_{\Omega\setminus\Gamma_C} \mathbb{D}e(\dot{u}) : e(\dot{u})\, \mathrm{d}x\, \mathrm{d}\tau + \mathrm{Var}_{\mathcal{R}}(0, t; z) \tag{11}$$

$$= \mathcal{E}(0, u(0), z(0)) + \int_0^t \partial_\tau \mathcal{E}(\tau, u(\tau, z(\tau))\, \mathrm{d}\tau , \tag{12}$$

with $\mathrm{Var}_{\mathcal{R}}(0, t; z) := \sup_{\text{partitions of } [0,T]} \sum_i \mathcal{R}(z(t_{i+1}) - z(t_i))$ the total variation induced by $\mathcal{R}$, which defines an extended quasi-distance ('extended' because $\mathcal{R}(v) = \infty$ is possible, 'quasi', because, as in (5), $\mathcal{R}$ need not be symmetric). A pair (u, z) that satisfies a suitable weak formulation of the momentum balance together with semistability (10) and the energy-dissipation balance (12) is called a *semistable energetic solution* to the system defined by the tuple $(\mathbf{V}, \mathbf{W}, \mathbf{Z}, \mathcal{V}, \mathcal{K}, \mathcal{R}, \mathcal{E})$. Results for adhesive contact, combined with further physical phenomena, based on this notion of solution can be found e.g. in [43–46].

The notion of *semistable energetic solutions* for coupled rate-dependent/rate-independent systems was for the first time studied in [48] and analyzed in [47] in a general, abstract setting, allowing for non-convex energies and non-quadratic dissipation potentials $\mathcal{V}$ being convex and of general superlinear growth. Then, in general, (12) holds only as an inequality '$\leq$' with $\int_0^t \int_{\Omega\setminus\Gamma_C} \mathbb{D}e(\dot{u}) : e(\dot{u})\, \mathrm{d}x\, \mathrm{d}\tau$ replaced by $\int_0^t \big(\mathcal{V}(\dot{u}(\tau)) + \mathcal{V}^*(\rho\ddot{u}(\tau) - \mathrm{D}_u\mathcal{E}(\tau, u(\tau), z(\tau)))\big)\, \mathrm{d}\tau$ for $\mathcal{V}^*$ the convex conjugate of $\mathcal{V}$. In [47, 48] it is also proven that semistable energetic solutions to a system $(\mathbf{V}, \mathbf{W}, \mathbf{Z}, \mathcal{V}, \mathcal{K}, \mathcal{R}, \mathcal{E})$ can be obtained using a time-discrete scheme, alternating between u and z at each time-step. Indeed, also in the fully rate-independent case, i.e., when all rate-dependent terms are eliminated from the momentum balance so that the evolution of u is quasistatic, such an alternate time-discrete scheme, results in a similar notion of solution that combines the semistability inequality (10) and an energy-dissipation inequality with the condition that $u(t)$ be a minimizer for $\mathcal{E}(t, \cdot, z(t))$, see e.g. [50] in the setting of adhesive contact. In fact, it has also been observed in [32, 52] that a vanishing-viscosity-and-inertia limit for a system $(\mathbf{V}, \mathbf{W}, \mathbf{Z}, \mathcal{V}, \mathcal{K}, \mathcal{R}, \mathcal{E})$ results in a semistable energetic formulation of the corresponding rate-independent, quasistatic limit system. Instead, if one uses in the rate-independent, quasistatic setting a monolithic time-discrete scheme, which simultaneously minimizes $\mathcal{E}(t_i, \cdot, \cdot) + \mathcal{R}(\cdot - z_{i-1})$ in the pair (u, z), then one arrives at the well-studied notion of *energetic solutions*, which consists of a quasistatic energy-dissipation balance akin to (12) and of a global *stability condition*, i.e., $\mathcal{E}(t, u(t), z(t) \leq \mathcal{E}(t, \tilde{u}, \tilde{z}) + \mathcal{R}(\tilde{z} - z(t))$ for all test functions $(\tilde{u}, \tilde{z})$. See e.g.

[36, 37] for abstract results on *energetic solutions* and [31] for adhesive contact. Energetic and semistable energetic solutions must not coincide: It is shown e.g. in [50] that energetic solutions tend to jump much earlier than semistable energetic solutions.

3 Adhesive Contact, Brittle, and Cohesive Zone-Type Delamination

The adhesive contact surface energy $\mathscr{E}_{\mathrm{adh}}$ from (1) allows for jumps $[\![u]\!]$ of the displacement field even in points of the interface where the bonds of the adhesive are not completely broken, i.e., $z \in (0,1)$. This may apply to a glue which allows the two parts of the body to slightly detach from each other, but not to a brittle material. In particular, brittle Griffith-type delamination, where the only surface energy is contributed by the energy dissipated due to crack-growth, cannot be modeled via $\mathscr{E}_{\mathrm{adh}}$ as long as the adhesive contact term $\int_{\Gamma_{\mathrm{C}}} \frac{k}{2} z \big|[\![u]\!]\big|^2 \,\mathrm{d}\mathscr{H}^{d-1}$ is involved. Delamination in a brittle material can rather be described by a surface energy involving the *brittle constraint* $I_{[z|[\![u]\!]|=0]}([\![u]\!], z)$, being $I_{[z|[\![u]\!]|=0]}([\![u]\!], z) = 0$ if $z|[\![u]\!]| = 0$ and $I_{[z|[\![u]\!]|=0]}([\![u]\!], z) = \infty$ otherwise. This constraint ensures that the displacements are continuous across Γ_{C} in points where no rupture has occurred yet, i.e., $z(t,x) \in (0,1]$ and allows the displacements to jump across the crack set $C(t) = \{x \in \Gamma_{\mathrm{C}},\, z(t,x) = 0\}$. Denoting by $\mathscr{E}_{\mathrm{brittle}}$ the surface energy given as in (1), but with the adhesive contact term $\frac{k}{2} z \big|[\![u]\!]\big|^2$ replaced by the brittle constraint $I_{[z|[\![u]\!]|=0]}([\![u]\!], z)$, it looks intuitive that $\mathscr{E}_{\mathrm{adh}}$ approximates $\mathscr{E}_{\mathrm{brittle}}$ as $k \to \infty$. In this spirit it was shown in [49], in the quasistatic, rate-independent setting that *energetic solutions of an adhesive contact problem* approximate an *energetic solution of a brittle delamination problem* as $k \to \infty$. Analogous approximation results for *semistable energetic solutions* in the *quasistatic, rate-independent setting*, resp. in the *coupled rate-dependent/rate-independent setting* have been obtained in [50], resp. in [45] featuring non-penetration but $\rho = 0$ in (6a), and in [46] for $\rho > 0$ but neglecting non-penetration. Indeed, in the coupled rate-dependent/rate-independent setting the main challenge lies in the limit passage in the weak formulation of the momentum balance. For $z : [0,T] \times \Gamma_{\mathrm{C}} \to [0,1]$ satisfying semistability (10) the brittle constraint imposes a non-smooth constraint (alike the non-penetration condition), but which additionally *depends on the properties of* z. This is a clear increase of difficulty, because suitable test functions v_k have to be tailored to the properties of z_k, in such a way as to, on the one hand, prevent in the weak formulation of the momentum balance a blow-up of the term $\int_{\Gamma_{\mathrm{C}}} k z_k(t)[\![u_k(t)]\!][\![v_k]\!] \,\mathrm{d}\mathscr{H}^{d-1}$ and, on the other hand, to recover all the information on the crack set of $z(t)$, being the semistable limit function at time $t \in [0,T]$ of the solutions $(z_k(t))_k$ of the adhesive problems. In particular [46] shows that the evolution of rate-independent brittle Griffith-type delamination in a visco-elastic solid with dynamic effects, can be described in terms of a weak form of the momentum balance, semistability (10), and an energy-dissipation balance. Thus, it contributes to the recent developments in the analytical understanding of dynamic fracture [18–20].

Alike adhesive contact, also cohesive zone models, pioneered e.g. in [6, 22, 42], regard fracture as a gradual phenomenon in which separation takes place across an extended crack 'tip', or cohesive zone. Thus, they also contain finite surface energy terms involving the displacement jump across $\Gamma_{\rm C}$ and, possibly an internal variable. Yet, therein, the internal variable ζ rather has the role to keep track of the history of the maximal opening displacement in normal direction. Thus, if ideally $\zeta(t,x) := \sup_{s\in[0,t]} [\![u(t,x)]\!]_{\mathbf{n}}$, it always has to hold $[\![u]\!]_{\mathbf{n}} \le \zeta$. This can be enforced by the indicator $I_{[0,\zeta(t,x)]}$ of the interval $[0,\zeta(t,x)]$. Inspired by [41], to allow for a different loading and unloading behavior of the surface energy, [57] considers

$$\mathscr{E}_{\rm coh}(u,\zeta) := \int_{\Gamma_{\rm C}} \Big(\frac{\phi_{\rm coh}(\zeta)}{2\zeta^2}\big|[\![u]\!]_{\mathbf{n}}\big|^2 + I_{[0,\zeta]}\big([\![u]\!]_{\mathbf{n}}\big) + I_{[0,\zeta^*]}(\zeta) + G(\zeta)\Big)\,\mathrm{d}\mathscr{H}^{d-1} \tag{13}$$

with $\phi_{\rm coh}$ a typical cohesive energy as proposed in [41] and G a gradient term (quadratic or Sobolev-Slobodeckij-type). Due to its role, ζ monotonously increases in time. The indicator $I_{[0,\zeta^*]}$ then ensures that the model is meaningful as long as the maximum opening displacements do not exceed an utter maximum ζ^*. [57] shows the existence of semistable energetic solutions for this model in the coupled rate-dependent/rate-independent setting (without inertia); to accomodate the increasing nature of ζ, the rate-independent dissipation is of the form $R(v) := a|v| + I_{[0,\infty)}(v)$.

While [41] and consequently [57] consider the cohesive term and the constraint $I_{[0,\zeta(t,x)]}$, keeping track of the opening history, as a part of the surface *energy*, many other analytical works consider them as part of the dissipation potential. In this setup, in [21], existence of energetic solutions in the rate-independent, quasistatic setting was shown without any gradient term G, thus calling for a different notion of convergence to overcome compactness issues. More recently, vanishing viscosity techniques have been applied in cohesive zone models with respect to the internal variable, [2, 14], the first also accounting for history dependence of the crack opening in a Young-measure setting. [5, 13] study cohesive zone delamination for a visco-elastic solid without introducing an internal variable and prove existence of solutions as well as a vanishing viscosity limit.

Acknowledgements The research of the author has been partially funded by the DFG (German Research Foundation) within Project *Finite element approximation of functions of bounded variation and application to models of damage, fracture, and plasticity* of the DFG Priority Programme SPP 1748 *Reliable Simulation Techniques in Solid Mechanics. Development of Non-standard Discretisation Methods, Mechanical and Mathematical Analysis*. This work was composed in the course of the International Conference *CoMFoS16 Mathematical Analysis of Continuum Mechanics and Industrial Applications II* held 2016 October 22th–24th at Kyushu University, Fukuoka, Japan. The author warmly thanks the organizing committee and, in particular, the organizers *Masato Kimura, Patrick van Meurs, and Hirofumi Notsu (all Kanazawa University)* for the invitation to the conference and for their hospitality at this successful event.

References

1. Akagi, G., Kimura, M.: Unidirectional evolution equations of diffusion type. arxiv:1501.01072 (2015)
2. Almi, S.: Energy release rate and quasistatic evolution via vanishing viscosity in a cohesive fracture model with an activation threshold. ESAIM Control Optim. Calc. Var. (2016). Published online
3. Ambati, M., Kruse, R., De Lorenzis, L.: A phase-field model for ductile fracture at finite strains and its experimental verification. Comput. Mech. **57** (2016)
4. Ambrosio, L., Fusco, N., Pallara, D.: Functions of Bounded Variation and Free Discontinuity Problems. Oxford University Press, New York (2005)
5. Artina, M., Cagnetti, F., Fornasier, M., Solombrino, F.: Linearly constrained evolutions of critical points and an application to cohesive fractures. arXiv-Preprint no. 1508.02965 (2016)
6. Barenblatt, G.: The mathematical theory of equilibrium of cracks in brittle fracture. Adv. Appl. Mech. **7**, 55–129 (1962)
7. Bonetti, E., Bonfanti, G., Rossi, R.: Well-posedness and long-time behaviour for a model of contact with adhesion. Indiana Univ. Math. J. **56**, 2787–2820 (2007)
8. Bonetti, E., Bonfanti, G., Rossi, R.: Global existence for a contact problem with adhesion. Math. Meth. Appl. Sci. **31**, 1029–1064 (2008)
9. Bonetti, E., Bonfanti, G., Rossi, R.: Analysis of a temperature-dependent model for adhesive contact with friction. Phys. D **285**, 42–62 (2014)
10. Bonetti, E., Bonfanti, G., Rossi, R.: Modeling via the internal energy balance and analysis of adhesive contact with friction in thermoviscoeleasticity. Nonlinear Analysis Real World Appl. **22**, 473–507 (2015)
11. Bonetti, E., Rocca, E., Scala, R., Schimperna, G.: On the strongly damped wave equation with constraint. WIAS-Preprint 2094 (2015)
12. Burke, S., Ortner, C., Süli, E.: An adaptive finite element approximation of a variational model of brittle fracture. SIAM J. Numerical Analysis **48**(3), 980–1012 (2010)
13. Cagnetti, F.: A vanishing viscosity approach to fracture growth in a cohesive zone model with prescribed crack path. Math. Models Methods Appl. Sci. **18**(7), 1027–1071 (2008)
14. Cagnetti, F., Toader, R.: Quasistatic crack evolution for a cohesive zone model with different response to loading and unloading: a young measures approach. ESAIM Control Optim. Calc. Var. **17**(1), 1–27 (2011)
15. Conti, S., Focardi, M., Iurlano, F.: Phase field approximation of cohesive fracture models. Ann. Inst. H. Poincaré Anal. Non Linéaire **33**(4), 1033–1067 (2016)
16. Conti, S., Focardi, M., Iurlano, F.: Some recent results on the convergence of damage to fracture. Atti Accad. Naz. Lincei Rend. Lincei Mat. Appl. **27**(1), 51–60 (2016)
17. Dal Maso, G.: Generalised functions of bounded deformation. J. Eur. Math. Soc. **15**, 1943–1997 (2013)
18. Dal Maso, G., Larsen, C.: Existence for wave equations on domains with arbitrary growing cracks. Atti Accad. Naz. Lincei Cl. Sci. Fis. Mat. Natur. Rend. Lincei (9) Mat. Appl. **22**, 387–408 (2011)
19. Dal Maso, G., Larsen, C., Toader, R.: Existence for constrained dynamic Griffith fracture with a weak maximal dissipation condition (2015). SISSA Preprint, Trieste
20. Dal Maso, G., Lazzaroni, G., Nardini, L.: Existence and uniqueness of dynamic evolutions for a peeling test in dimension one (2016). SISSA Preprint, Trieste
21. Dal Maso, G., Zanini, C.: Quasi-static crack growth for a cohesive zone model with prescribed crack path. Proc. Roy. Soc. Edinb. Sect. A **137**(2), 253–279 (2007)
22. Dugdale, D.: Yielding of steel sheets containing clits. J. Mech. Phys. Solids **8**, 100–104 (1960)
23. Francfort, G.A., Marigo, J.J.: Revisiting brittle fracture as an energy minimization problem. J. Mech. Phys. Solids **46**(8), 1319–1342 (1998)
24. Freddi, F., Iurlano, F.: Numerical insight of a variational smeared approach to cohesive fracture. J. Mech. Phys. Solids **98**, 156–171 (2017)

25. Frémond, M.: Contact with adhesion. In: Moreau, J., Panagiotopoulos, P., Strang, G. (eds.) Topics in Nonsmooth Mechanics, pp. 157–186. Birkhäuser (1988)
26. Frémond, M.: Non-Smooth Thermomechanics. Springer, Heidelberg (2002)
27. Giacomini, A.: Ambrosio-Tortorelli approximation of quasi-static evolution of brittle fracture. Calc. Var. Partial Differential Equations **22**, 129–172 (2005)
28. Giacomini, A., Ponsiglione, M.: Discontinuous finite element approximation of quasistatic crack growth in finite elasticity. Math. Models Methods Appl. Sci. **16**(1), 77–118 (2006)
29. Halphen, B., Nguyen, Q.: Sur les matériaux standards généralisés. J. Mécanique **14**, 39–63 (1975)
30. Keip, M.A., Kiefer, B., Schröder, J., Linder, C. (eds.): Special issue on phase field approaches to fracture: in memory of Professor Christian Miehe (19562016). Comput. Methods Appl. Mech. Eng. **312** (2016)
31. Kočvara, M., Mielke, A., Roubíček, T.: A rate-independent approach to the delamination problem. Math. Mech. Solids **11**, 423–447 (2006)
32. Lazzaroni, G., Rossi, R., Thomas, M., Toader, R.: Rate-independent damage in thermo-viscoelastic materials with inertia (2014). WIAS Preprint 2025
33. Marigo, J., Maurini, C., Pham, K.: An overview of the modelling of fracture by gradient damage models. Meccanica **51**(12), 3107–3128 (2016)
34. Miehe, C., Hofacker, M., Welschinger, F.: A phase field model for rate-independent crack propagation: robust algorithmic implementation based on operator splits. Comput. Methods Appl. Mech. Eng. (2010)
35. Miehe, C., Welschinger, F., Hofacker, M.: Thermodynamically consistent phase-field models of fracture: variational principles and multi-field FE implementations. Int. J. Numer. Meth. Engng. **83**, 12731311 (2010)
36. Mielke, A.: Evolution in rate-independent systems (Chap. 6). In: C. Dafermos, E. Feireisl (eds.) Handbook of Differential Equations, Evolutionary Equations, vol. 2, pp. 461–559. Elsevier B.V., Amsterdam (2005)
37. Mielke, A., Roubíček, T.: Rate-Independent Systems: Theory and Application. Applied Mathematical Sciences. Springer, New York (2015)
38. Mielke, A., Theil, F.: On rate-independent hysteresis models. NoDEA Nonlinear Differential Equations Appl. **11**(2), 151–189 (2004)
39. Mielke, A., Theil, F., Levitas, V.I.: A variational formulation of rate-independent phase transformations using an extremum principle. Arch. Ration. Mech. Anal. **162**, 137–177 (2002)
40. Ohtsuka, K.: Comparison of criteria on the direction of crack extension. J. Comput. Appl. Math. **149**, 335–339 (2002)
41. Ortiz, M., Pandolfi, A.: Finite-deformation irreversible cohesive elements for three-dimensional crack-propagation analysis. Int. J. Numer. Meth. Eng. **44**, 1267–1282 (1999)
42. Rice, J.: Fracture, Chapter Mathematical Analysis in the Mechanics of Fracture, pp. 191–311. Academic Press, New York (1968)
43. Rossi, R., Roubíček, T.: Thermodynamics and analysis of rate-independent adhesive contact at small strains. Nonlinear Anal. **74**(10), 3159–3190 (2011)
44. Rossi, R., Roubíček, T.: Adhesive contact delaminating at mixed mode, its thermodynamics and analysis. Interfaces Free Bound. **15**(1), 1–37 (2013)
45. Rossi, R., Thomas, M.: From an adhesive to a brittle delamination model in thermo-visco-elasticity. ESAIM Control Optim. Calc. Var. **21**, 1–59 (2015)
46. Rossi, R., Thomas, M.: From adhesive to brittle delamination in visco-elastodynamics. WIAS-Preprint 2259 (2016)
47. Rossi, R., Thomas, M.: Coupling rate-independent and rate-dependent processes: existence results. WIAS-Preprint 2123. SIMA, accepted (2017)
48. Roubíček, T.: Rate-independent processes in viscous solids at small strains. Math. Methods Appl. Sci. **32**(7), 825–862 (2009)
49. Roubíček, T., Scardia, L., Zanini, C.: Quasistatic delamination problem. Continuum Mech. Thermodynam. **21**(3), 223–235 (2009)

50. Roubíček, T., Thomas, M., Panagiotopoulos, C.: Stress-driven local-solution approach to quasistatic brittle delamination. Nonlinear Anal. Real World Appl. **22**, 645–663 (2015)
51. Scala, R.: A weak formulation for a rate-independent delamination evolution with inertial and viscosity effects subjected to unilateral constraint. WIAS-Preprint 2172 (2015)
52. Scala, R.: Limit of viscous dynamic processes in delamination as the viscosity and inertia vanish. ESAIM COCV **23**, 593–625 (2017)
53. Scala, R., Schimperna, G.: A contact problem for viscoelastic bodies with inertial effects and unilateral boundary constraints,. WIAS-Preprint 2147 (2015)
54. Schlüter, A., Willenbächer, A., Kuhn, C., Müller, R.: Phase field approximation of dynamic brittle fracture. Comput. Mech. **54**, 1141–1161 (2014)
55. Takaishi, T., Kimura, M.: Phase field model for mode III crack growth in two dimensional elasticity. Kybernetika **45**(4), 605–614 (2009)
56. Thomas, M., Mielke, A.: Damage of nonlinearly elastic materials at small strain: existence and regularity results. Zeit. Angew. Math. Mech. **90**(2), 88–112 (2010)
57. Thomas, M., Zanini, C.: Cohesive zone-type delamination in visco-elasticity. WIAS-preprint 2350 (2016)
58. Watanabe, K., Azegami, H.: Proposal of new stability-instability criterion for crack extension based on crack energy density and physical sytematization of other criteria. Bull. JSME **28**(246), 2873–2880 (1985)
59. Weinberg, K., Dally, T., Schuß, S., Werner, M., Bilgen, C.: Modeling and numerical simulation of crack growth and damage with a phase field approach. GAMM-Mitt. **39**(1), 55–77 (2016)

Simulation of Ductile Fracture in Amorphous and Polycrystalline Materials by Multiscale Cohesive Zone Model

Shingo Urata and Shaofan Li

Abstract A multiscale cohesive zone model (MCZM) that combines finite element method with atomistic modeling is applied to simulate fracture of amorphous materials and polycrystalline solids. In order to apply MCZM to model amorphous materials, the Cauchy–Born rule is linked with the Parrinello–Rahman MD method to associate atom configurations with material deformation by using molecular statics (MS). We found the algorithm allows us to simulate ductile fracture of amorphous materials successfully. In addition, the methodology is applied to model the amorphous grain boundaries of polycrystalline solids, and we show that it can capture ductile fracture of polycrystalline metals.

1 Introduction

Since fracture and cracking are common and crucial damages for brittle materials in industrial applications, much effort has been devoted by the research community to develop practical computational techniques that can simulate such irreversible physical process with broad engineering applications.

Extended finite element method (XFEM) [1, 2] and cohesive zone model (CZM) [3, 4] are now available in most of commercial finite element softwares, and they have been widely applied to simulate discontinuous material damage in failure analysis. XFEM utilizes the Heaviside function to represent discontinuous displacement field along crack surfaces. On the other hand, CZM assumes zero thickness

S. Urata
Innovative Technology Research Center, Asahi Glass Co. Ltd., 1150 Hazawa-cho, Kanagawa-ku, Yokohama-Shi, Kanagawa 221-8755, Japan
e-mail: shingo-urata@agc.com

S. Li (✉)
Department of Civil and Environmental Engineering, University of California, Berkeley, CA 94720, USA
e-mail: shaofan@berkeley.edu

P. van Meurs et al. (eds.), *Mathematical Analysis of Continuum Mechanics and Industrial Applications II*, Mathematics for Industry 30,
https://doi.org/10.1007/978-981-10-6283-4_4

interface between two finite elements, to presume fracture process zone at the interface [5]. Other approaches such as meshfree particle method modeling of fracture [6], peridynamics method proposed by Silling [7, 8], the hybrid particle/finite element method (PDS-FEM) have also been developed to simulate fracture by using conjugate geometries of Voronoi and Delaunay tessellations [9–11]. Another important class of approaches is the phase-field modeling of fracture, in which a crack may be represented by an evolving phase-field level set function, e.g., [12–16]. Francfort and Marigo's variational approach has been adopted in the phase-field approach to model crack growth by considering energy associated with crack nucleation.

These continuum methods mentioned above require experimentally observed material properties, such as fracture toughness, Young's modulus and Shear modulus, to represent stiffness and crack criteria of materials in general. On the other hand, classical and ab-initio molecular dynamics simulations allow us to simulate crack propagation based on only empirical or theoretical interatomic interactions. However, it is still limited to apply these atomistic simulations to only nano-scale phenomena (e.g. [17, 18]), even though computational availability has been progressed continuously. It is therefore a variety of multiscale simulation technologies have been developed to conquer difficulty in scale length with combining atomistic and continuum simulation schemes. As mentioned in a review, the multiscale simulations are divided into two categories: concurrent and hierarchical approaches [19]. Multiscale cohesive zone mode (MCZM) proposed by Li et al. [20, 21] is one of the hierarchical multiscale modelings. The method allows us to simulate material response based on interatomic interaction among atoms in an unit cell, which is embedded into the each of every element that forms a discretization of material body that is under the consideration. The unique idea of MCZM is employing extremely thin element between bulk elements as a finite width cohesive zone or process zone. This allows us to apply the hierarchical Cauchy–Born rule method in the cohesive zone, by doing so we can capture macroscale material fracture by only considering nano-scale atomistic interaction. The method has been applied to simulate fracture of model crystalline solid [22], polycrystalline solids [23], for example copper [24]. In addition, we have implemented three body interaction models to examine single-crystalline silicon [25], and further extended it to study amorphous materials with combining molecular statics (MS) [26], which can minimize potential energy of atoms in an unit cell by conjugate-gradient method (CG).

In this paper, we first present an overview of MCZM models including recently developed algorithm for amorphous materials in Sect. 2. After demonstrating two numerical examples of fracture simulations in Sect. 3, we finally describe brief conclusions in Sect. 4.

2 Computational Materials Modeling

2.1 *Cohesive Zone Model*

Since the MCZM is a multiscale version of CZM, we briefly introduce conventional CZM at first. CZM is a relatively simple method to apply FEM procedure to simulate fracture without singularity problem. The method allows separation of finite elements which are bonded by an interface, namely fracture process zone, to represent fracture or crack propagation. Figure 1 draws schematic traction–separation relation at the interface, where stress is induced as a function of distance between two facets of bulk elements. When a crack is initiated, stress reaches maximum (σ_{max}) and then the stress gradually decreases until the interface is separated completely. As mentioned in [5], although a variety of functions have been proposed to describe cohesive traction–separation relation there are basically two types of cohesive laws used in practice, i.e., the polynomial and the exponential types suggested by Needleman and his co-authors [27, 28]; main differences among these cohesive traction-displacement functions are their shape and parameters corresponding to different cohesive responses in different materials. Since shape of the cohesive function should be modified based on experimental observations, conventional CZM is basically a semi-empirical method of simulating fracture.

2.2 *Multiscale Cohesive Zone Model*

The finite element discrete equations of motion can be derived from Galerkin weak formulation as described in the literature, e.g., [20–25]. Here, we briefly outline the discrete equation of motion as follows,

$$\mathbf{M}\ddot{\mathbf{d}} + \mathbf{f}^{int}(\mathbf{d}) - \mathbf{f}^{cohe}(\mathbf{d}) = \mathbf{f}^{ext} \tag{1}$$

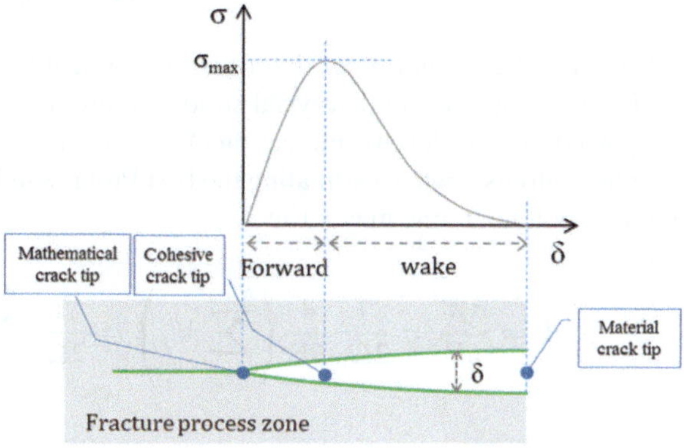

Fig. 1 An empirical cohesive traction–separation curve in fracture process zone

where, $\mathbf{M}$ and $\mathbf{f}^{ext}$ are the mass matrix and external force. $\mathbf{f}^{int}$ and $\mathbf{f}^{cohe}$ are force vectors from bulk elements and CZ elements. They are defined as follows:

$$\mathbf{f}^{int} = \mathop{\mathbf{A}}_{e=1}^{n_B^e} \int_{\Omega_B^e} \mathbf{B}^{eT} \mathbf{P}^e(\mathbf{d}) dV \tag{2}$$

$$\mathbf{f}^{cohe} = \mathop{\mathbf{A}}_{e=1}^{n_{CZ}^e} \int_{\Omega_{CZ}^e} \left\{ \mathbf{B}^{eT} \mathbf{P}^e(\mathbf{d}) + \mathbf{C}^{eT} \mathbf{Q}^e(\mathbf{d}) \right\} dV \tag{3}$$

where, $\mathbf{A}$ is the element assemble operator over number of elements for bulk (n_B^e) and CZ (n_{CZ}^e). $\mathbf{N}^e$, $\mathbf{B}^e$, and $\mathbf{C}^e$ are the element shape function, strain-displacement matrix and the gradient of the strain-displacement matrix, respectively. $\mathbf{P}$ and $\mathbf{Q}$ are the first Piola–Kirchhoff stress tensor (PK_I) and the second-order stress tensor. A specific feature of MCZM is applying atomistic model and their interatomic interaction to deduce the stress tensors as follows.

In order to construct the traction–separation relation by using interatomic interaction instead of empirical estimation according to experiments, an unit cell composed of multiple atoms is embedded into every finite elements including CZ elements [20, 21]. Here, we have to introduce an important assumption, Cauchy–Born rule (CBR), which postulates uniform deformation in each element and embedded atoms follow the deformation of the element [29]. In this case, current position of atom i ($\mathbf{r}_i$) can be easily evaluated from original position $\mathbf{R}_i$ through deformation gradient $\mathbf{F}$ of an element as,

$$\mathbf{r}_i = \mathbf{F}\mathbf{R}_i, \ \mathbf{F} = \frac{\partial \mathbf{x}}{\partial \mathbf{X}} = \mathbf{I} + \frac{\partial N}{\partial \mathbf{X}} \cdot \mathbf{d} \tag{4}$$

where N is shape function of element and $\mathbf{d}$ is nodal displacement. $\mathbf{x}$ and $\mathbf{X}$ are deformed and undeformed position vectors, respectively. Further, higher order CBR is available if we postulate higher order deformation [25, 30, 31]. For instance, if we take up to the second-order deformation, atom configuration can be expressed as;

$$\mathbf{r}_i = \mathbf{F}\mathbf{R}_i + \frac{1}{2}\mathbf{G} : (\mathbf{R}_i \otimes \mathbf{R}_i), \ \mathbf{G} = \frac{\partial \mathbf{F}}{\partial \mathbf{X}} = \frac{\partial^2 \mathbf{x}}{\partial \mathbf{X} \otimes \partial \mathbf{X}} = \frac{\partial^2 N}{\partial \mathbf{X} \otimes \partial \mathbf{X}} \cdot \mathbf{d} \tag{5}$$

where $\mathbf{G}$ is the second-order deformation gradient tensor.

For monoatomic single crystal solids, all atoms are identical in the unit cell, we only need to consider the interaction (V_{ij}) between the representative atom i and its neighbor atoms j when estimating the first Piola–Kirchoff stress (PK_I) tensor $\mathbf{P}$. For the pair potential, one may write

$$\mathbf{P} = \frac{\partial W}{\partial \mathbf{F}} = \frac{1}{2\Omega_0} \frac{\partial}{\partial \mathbf{F}} \left(\sum_{j=1}^{N_{atom}} V_{ij} \right) = \frac{1}{2\Omega_0} \sum_{j=1}^{N_{atom}} \frac{\partial V_{ij}}{\partial r_{ij}} \frac{\mathbf{r}_{ij} \otimes \mathbf{R}_{ij}}{r_{ij}}, \tag{6}$$

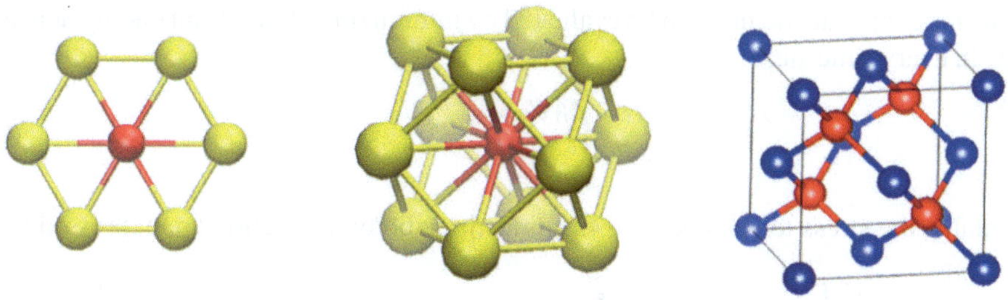

(a) Hexagonal unit cell in 2D (b) FCC unit cell in 3D (c) Cubic diamond lattice unit cell in 3D

Fig. 2 Typical unit cells for **a** 2D hexagonal lattice, **b** 3D FCC lattice, and **c** 3D cubic diamond lattice. For **a** and **b** red shows central atom and yellow indicates surrounding atoms. In **c**, red and blue atoms belong to different FCC lattices

where $\mathbf{r}_{ij} = \mathbf{r}_i - \mathbf{r}_j, r_{ij} = |\mathbf{r}_i - \mathbf{r}_j|$, and $\mathbf{R}_{ij} = \mathbf{R}_i - \mathbf{R}_j$. W is called as strain energy density and Ω_0 represents the volume of the unit cell in the initial configuration. N_{atom} is number of atoms in the unit cell. Similarly, the second-order stress tensor is,

$$\mathbf{Q} = \frac{\partial W}{\partial \mathbf{G}} = \frac{1}{2\Omega_0} \frac{\partial}{\partial \mathbf{G}} \left(\sum_{j=1}^{N_{atom}} V_{ij} \right) = \frac{1}{4\Omega_0} \sum_{j=1}^{N_{atom}} \left[\left(\frac{\partial V_{ij}}{\partial r_{ij}} \frac{1}{r_{ij}} \right) \mathbf{r}_{ij} \otimes \mathbf{R}_{ij} \otimes \mathbf{R}_{ij} \right] \tag{7}$$

Figure 2 shows some of typical configurations of unit cells of crystalline solids, which have been used in some of 2D simulation cases, e.g., [20, 21, 23, 30–32] and also in 3D simulations, e.g., [24, 25] of MCZM. If we consider single crystal solid in 2D with only the nearest neighbors taken into account, the unit cell is, for instance for FCC crystals, a hexagonal shape composed of one central atom and six surrounding atoms as shown in Fig. 2a. In Fig. 2b, we show the 3D FCC lattice unit cell. Figure 2c shows a more complicated lattice unit cell, diamond cubic, which can be represented as a pair of FCC lattices (Fig. 2c). In this paper, we would like to simulate ductile fracture of amorphous silicon, and we choose the Tersoff potential [33] in multiscale process zone model, which may be used to quantify atomistic interactions for both crystalline silicon (c-Si) and amorphous silicon (a-Si). The Tersoff potential is a three body interaction potential, and it takes only the nearest neighbors into account to evaluate interatomic interaction. For crystalline silicon solids, we only need one atom in one FCC lattice unit cell and four surrounding atoms belonging to another FCC lattice unit cell.

Note that because diamond cubic lattice has an additional freedom to represent distance between two FCC lattices ($\mathbf{v}$), PK_I of silicon can be written as,

$$\mathbf{P} = \frac{\partial W}{\partial \mathbf{F}} = \frac{1}{2\Omega_0} \frac{\partial}{\partial \mathbf{F}} \left(\sum_{j=2}^{5} V_{ij} \right) = \frac{1}{2\Omega_0} \sum_{j=2}^{5} \left(\frac{\partial V_{ij}}{\partial \mathbf{F}} + \frac{\partial V_{ij}}{\partial \mathbf{v}} \frac{\partial \mathbf{v}}{\partial \mathbf{F}} \right) . \tag{8}$$

The inner displacement **v** can be evaluated by minimizing of the strain energy density W in every time steps,

$$\left(\frac{\partial V_{ij}}{\partial \mathbf{v}}\right)_{\mathbf{F}} = 0\,. \tag{9}$$

Finally, the stress tensors derived from the three body interaction are expressed as

$$\mathbf{P} = \frac{1}{2\Omega_0}\sum_{j=2}^{5}\left[\frac{\partial V_{1j}}{\partial r_{1j}}\frac{\partial r_{1j}}{\partial \mathbf{F}} + \sum_{k=2,k\neq j}^{5}\left(\frac{\partial V_{1j}}{\partial r_{1k}}\frac{\partial r_{1k}}{\partial \mathbf{F}} + \frac{\partial V_{1j}}{\partial cos\theta_{1jk}}\frac{\partial cos\theta_{1jk}}{\partial \mathbf{F}}\right)\right] \tag{10}$$

$$\mathbf{Q} = \frac{1}{2\Omega_0}\sum_{j=2}^{5}\left[\frac{\partial V_{1j}}{\partial r_{1j}}\frac{\partial r_{1j}}{\partial \mathbf{G}} + \sum_{k=2,k\neq j}^{5}\left(\frac{\partial V_{1j}}{\partial r_{1k}}\frac{\partial r_{1k}}{\partial \mathbf{G}} + \frac{\partial V_{1j}}{\partial cos\theta_{1jk}}\frac{\partial cos\theta_{1jk}}{\partial \mathbf{G}}\right)\right]. \tag{11}$$

Readers may consult more detail in [25].

It is remarkable that shown in Eq. (3) that the force vector in CZ element contains the second-order stress tensor that means that higher order CBR is taken into consideration in CZ element. This is because higher order deformation elongates atom distances, and as a result it induces weakness of CZ element compared with bulk element. Because cohesive zone is the fracture process zone as previously explained, it should be intrinsically weaker than bulk element. In order to explain the fragility, several ideas have been examined. For example, interface depletion potential [21, 23], asymptotic and mixed atomistic potential [22], and higher order CBR [25, 30, 31] have been successfully applied.

We have found nonlinear deformation represented by bubble node is also applicable to deduce the potential in CZ element without using higher order CBR [25]. In this method, inserting a bubble node in the middle of CZ element (wedge element) and the position is associated with scale of deformation of the element through the sigmoidal function.

$$\zeta_{bub}(t) = \frac{2}{1 + exp\{-P_{nl}(|\mathbf{F}(t-1)\cdot\mathscr{N}| - 1)\}} - 1, \tag{12}$$

where, ζ_{bub} is the normalized coordinate of bubble node on ζ direction at time t. $\mathscr{N}$ is the normal vector of the CZ element, and $\mathbf{F}(t-1)$ is the first order deformation gradient at previous time step. Since $|\mathbf{F}\cdot\mathscr{N}|$ represents degree of deformation in the direction of cohesive zone element thickness, more nonlinear deformation is evoked at more deformed condition. Parameter P_{nl} in Eq. (12) can adjust the sensitivity of the bubble node position.

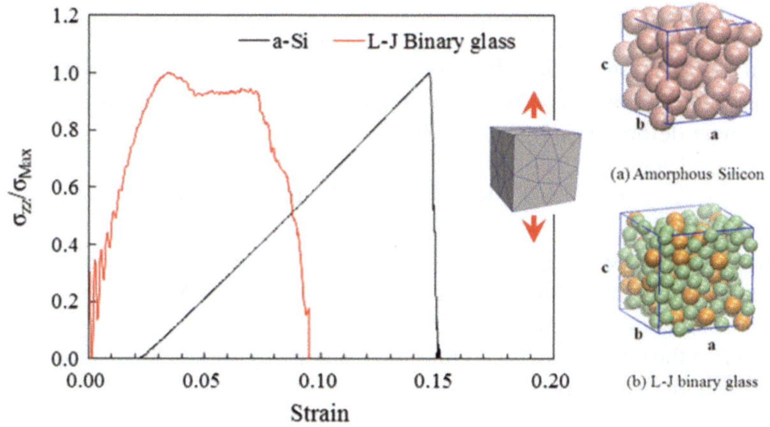

Fig. 3 Stress–strain curves of amorphous silicon and Lennard-Jones binary glass with uniaxial stretch. Representative cell for **a** amorphous silicon with 64 atoms and **b** Lennard-Johns binary glass with 256 atoms

2.3 *Modeling for Amorphous Materials*

Since the amorphous materials don't have lattice structure, it is necessary to define certain size of polyatomic representative cell (r-cell). In our previous study [26], minimum size of r-cells, which contains 64 and 256 atoms for a-Si and LJBG, respectively, are applied (see Fig. 3), since it has been found that the minimum cells can reasonably reproduce materials response. In order to develop a CBR procedure in amorphous materials, we consider the following algorithm, which is called as coarse-grained Parrinello–Rahman method (CG-PR) [26], by borrowing an idea of Parrinello–Rahman molecular dynamics [34, 35]. We may define a unit cell shape tensor as

$$\mathbf{h}(0) = [\mathbf{a}, \mathbf{b}, \mathbf{c}]. \tag{13}$$

According to [35], we may use the cell shape tensor $\mathbf{h}$ to re-scale the initial positions of atoms, i.e.,

$$\mathbf{R}_i = \mathbf{h}(0)\mathbf{S}_i, \quad i = 1, 2, \cdots N_c \,, \tag{14}$$

where $\mathbf{S}_i$ is the scaled atom positions (see [34] for discussions). Assuming an affine mapping $\boldsymbol{\varphi} : \mathbf{R} \to \mathbf{r}$, the deformation gradient may be defined as $\mathbf{F} = \dfrac{\partial \boldsymbol{\varphi}}{\partial \mathbf{R}}$. Then we can write down the Cauchy–Born rule as,

$$\mathbf{r}_i = \mathbf{F}\mathbf{R}_i = \mathbf{F} \cdot \mathbf{h}(0) \cdot \mathbf{S}_i = \mathbf{h}(t)\mathbf{S}_i \,, \tag{15}$$

where $\mathbf{h}(t) = \mathbf{F} \cdot \mathbf{h}(0)$.

In the CG-PR, next step atom configurations can be computed as follows:

1. Find deformation gradient $\mathbf{F}(t)$ for each element by using FEM procedure.
2. Guess next step atom configurations $\mathbf{r}_{gi}(t)$ from previous coordinates with assuming linear deformation by using CBR.

$$\mathbf{r}_{gi}(t) = \mathbf{F} \cdot \mathbf{R}_i(t - \Delta t) = \mathbf{F} \cdot \mathbf{h}(0) \cdot \mathbf{S}_i(t - \Delta t), \tag{16}$$

   where, $\mathbf{h}(0)$ is the initial r-cell shape tensor, $\mathbf{R}_i(t - \Delta t)$ is atom coordinate in referential configuration at previous time step, and $\mathbf{S}_i(t - \Delta t)$ is normalized coordinates of atoms in the r-cell.
3. Optimize the configurations to minimize energy by using CG method with periodic boundary condition in all direction.

$$\mathbf{r}_{opti}(t) = \arg\min V(\mathbf{r}_i) \tag{17}$$

4. Find current scaled atom positions $\mathbf{S}_i(t)$.

Once optimized configuration $\mathbf{r}_{opt}$ is found, we can calculate the first Piola–Kirchhoff stress tensor $\mathbf{P}$ by Eqs. (6) and (8). Note that above algorithm does not consider the second-order deformation $\mathbf{G}$, because CZ element may not support high-order strain fields, or strain gradients in Eq. (3), unless higher order elements are used. However, by applying the sigmoidal function Eq. (12), we may recover some nonlinear deformation in CZ elements in order to model fracture process zone.

Note here that the MCZM method combined with the CG-PR procedure doesn't take temperature effect in essence, because atom configurations are statistically optimized. Further development is required to consider thermal effect as discussed in [32].

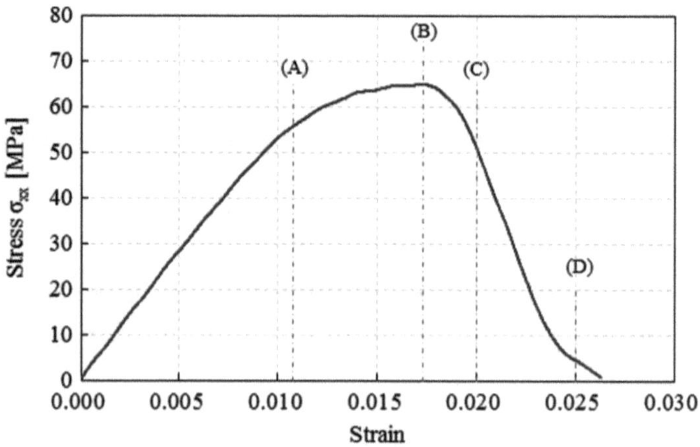

Fig. 4 Stress–strain relation for model polycrystalline material composed of Lennard-Jones particles. (A) Crack initiation (B) maximum stress, (C) 0.02 strain, and (D) complete separation

3 Numerical Examples

Uniaxial stretch of model amorphous materials

In order to confirm applicability of our algorithm mentioned in Sect. 2.3, fracture simulation of a cubic specimen was conducted for a-Si and LJBG with applying uniaxial stretch. It is seen in Fig. 3 that MCZM can capture both brittle and ductile fracture based on atomistic configurations and their interactions. More detail is discussed in [26].

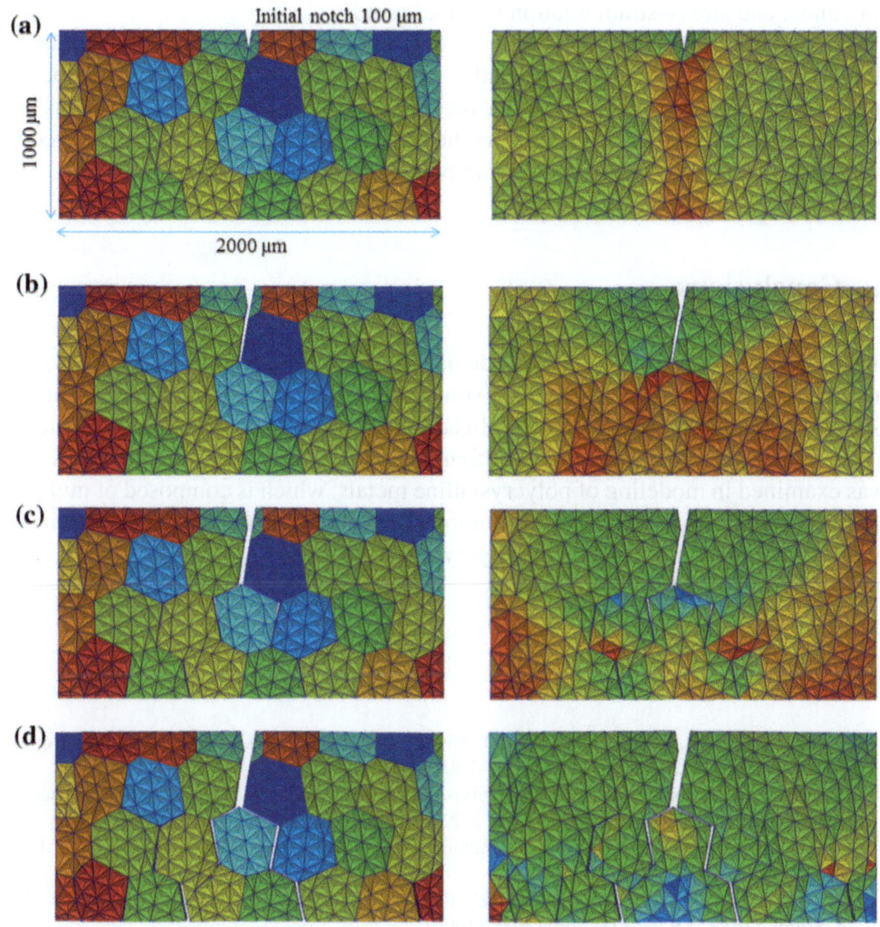

Fig. 5 Crack path and (Right) stress distribution during crack propagation. Indexes **a**–**d** correspond to strain shown in Fig. 4

Model polycrystalline material

A schematic view and size of the single-notched polycrystalline material are depicted in Fig. 4. In order to represent atomistic configuration at grain boundary, we utilize the amorphous r-cell of LJBG (see Fig. 3a) for only CZ element at the boundary, because amorphous phase is usually seen in grain boundaries [36, 37]. Lennard-Jones potential parameters for the two particles A and B in the r-cell are $\varepsilon_{AA} = 1.0\varepsilon_0$, $\varepsilon_{BB} = 1.5\varepsilon_0$, $\varepsilon_{AB} = 0.5\varepsilon_0$, $\sigma_{AA} = 1.0\sigma_0$, $\sigma_{BB} = 0.8\sigma_0$ and $\sigma_{AB} = 0.88\sigma_0$, respectively. Whereas for FCC crystal lattice (Fig. 2b), the L-J type potential for particle A was assumed in bulk elements, and CZ elements are embedded inside of grains. Constant velocity was applied as boundary condition at both sides of the specimen, and stress–strain relation was measured.

Figure 4 displays stress profile during stretching the specimen in horizontal direction. The stress increases monotonically as elastic deformation grows until crack is initiated at about 1.1% strain. After that the stress softening is observed, stress reaches to maximum at around 1.7% strain. As the corresponding snapshots are depicted in Fig. 5, the crack path through only grain boundary.

4 Conclusions

In this paper, we have discussed fundamental aspects of multiscale cohesive zone model, which have been developed to simulate fracture of materials based on atomistic interaction with crystal unit cell. In addition, recently developed computational algorithm to simulate amorphous configurations with assuming Cauchy–Born rule was examined in modeling of polycrystalline metals, which is composed of multiple grains. It has shown that by applying amorphous representative cell to grain boundary we can capture ductile fracture of polycrystalline metals.

References

1. Dolbow, J.O.H.N., Belytschko, T.: A finite element method for crack growth without remeshing. Int. J. Numer. Methods Eng. **46**(1), 131–150 (1999)
2. Stolarska, M., Chopp, D.L., Mos, N., Belytschko, T.: Modelling crack growth by level sets in the extended finite element method. Int. J. Numer. Methods Eng. **51**(8), 943–960 (2001)
3. Xu, X.P., Needleman, A.: Numerical simulations of fast crack growth in brittle solids. J. Mech. Phys. Solids **42**(9), 1397–1434 (1994)
4. Camacho, G.T., Ortiz, M.: Computational modelling of impact damage in brittle materials. Int. J. Solids Struct. **33**(20–22), 2899–2938 (1996)
5. Shet, C., Chandra, N.: Analysis of energy balance when using cohesive zone models to simulate fracture processes. J. Eng. Mater. Technol. **124**(4), 440–450 (2002)
6. Ren, B., Li, S.: Modeling and simulation of large-scale ductile fracture in plates and shells. Int. J. Solids Struct. **49**(18), 2373–2393 (2012)
7. Silling, S.A.: Reformulation of elasticity theory for discontinuities and long-range forces. J. Mech. Phys. Solids **48**(1), 175–209 (2000)

8. Silling, S.A., Lehoucq, R.B.: Peridynamic theory of solid mechanics. Adv. Appl. Mech. **44**, 73–168 (2010)
9. Hori, M., Oguni, K., Sakaguchi, H.: Proposal of FEM implemented with particle discretization for analysis of failure phenomena. J. Mech. Phys. Solids **53**(3), 681–703 (2005)
10. Oguni, K., Wijerathne, M.L.L., Okinaka, T., Hori, M.: Crack propagation analysis using PDS-FEM and comparison with fracture experiment. Mech. Mater. **41**(11), 1242–1252 (2009)
11. Wijerathne, M.L.L., Oguni, K., Hori, M.: Numerical analysis of growing crack problems using particle discretization scheme. Int. J. Numer. Methods Eng. **80**(1), 46–73 (2009)
12. Francfort, G.A., Marigo, J.J.: Revisiting brittle fracture as an energy minimization problem. J. Mech. Phys. Solids **46**(8), 1319–1342 (1998)
13. Bourdin, B., Francfort, G.A., Marigo, J.J.: Numerical experiments in revisited brittle fracture. J. Mech. Phys. Solids **48**(4), 797–826 (2000)
14. Bourdin, B.: Numerical implementation of the variational formulation for quasi-static brittle fracture. Interfaces Free Bound. **9**(3), 411–430 (2007)
15. Bourdin, B., Francfort, G.A., Marigo, J.J.: The variational approach to fracture. J. Elast. **91**(1–3), 5–148 (2008)
16. Takaishi, T., Kimura, M.: Phase field model for mode III crack growth in two dimensional elasticity. Kybernetika **45**(4), 605–614 (2009)
17. Rountree, C.L., Kalia, R.K., Lidorikis, E., Nakano, A., Van Brutzel, L., Vashishta, P.: Atomistic aspects of crack propagation in brittle materials: multimillion atom molecular dynamics simulations. Ann. Rev. Mater. Res. **32**(1), 377–400 (2002)
18. Goel, S., Faisal, N.H., Luo, X., Yan, J., Agrawal, A.: Nanoindentation of polysilicon and single crystal silicon: molecular dynamics simulation and experimental validation. J. Phys. D: Appl. Phys. **47**(27), 275304 (2014)
19. Miller, R.E., Tadmor, E.B.: A unified framework and performance benchmark of fourteen multiscale atomistic/continuum coupling methods. Model. Simul. Mater. Sci. Eng. **17**(5), 053001 (2009)
20. Liu, X., Li, S., Sheng, N.: A cohesive finite element for quasi-continua. Comput. Mech. **42**(4), 543–553 (2008)
21. Zeng, X., Li, S.: A multiscale cohesive zone model and simulations of fractures. Comput. Methods Appl. Mech. Eng. **199**(9), 547–556 (2010)
22. Li, S., Zeng, X., Ren, B., Qian, J., Zhang, J., Jha, A.K.: An atomistic-based interphase zone model for crystalline solids. Comput. Methods Appl. Mech. Eng. **229**, 87–109 (2012)
23. Qian, J., Li, S.: Application of multiscale cohesive zone model to simulate fracture in polycrystalline solids. J. Eng. Mater. Technol. **133**(1), 011010 (2011)
24. He, M., Li, S.: An embedded atom hyperelastic constitutive model and multiscale cohesive finite element method. Comput. Mech. **49**(3), 337–355 (2012)
25. Urata, S., Li, S.: Higher order Cauchy–Born rule based multiscale cohesive zone model and prediction of fracture toughness of silicon thin films. Int. J. Fract. **203**, 159–181 (2017)
26. Urata, S., Li, S.: A multiscale model for amorphous materials. Comput. Mater. Sci. **135**, 64–77 (2017)
27. Needleman, A.: A continuum model for void nucleation by inclusion debonding. J. Appl. Mech. **54**(3), 525–531 (1987)
28. Needleman, A.: An analysis of decohesion along an imperfect interface. Int. J. Fract. **42**(1), 21–40 (1990)
29. Ericksen, J.L.: On the Cauchy–Born rule. Math. Mech. Solids **13**(3–4), 199–220 (2008)
30. Fan, H., Li, S.: Multiscale cohesive zone modeling of crack propagations in polycrystalline solids. GAMM-Mitteilungen **38**(2), 268–284 (2015)
31. Lyu, D., Fan, H., Li, S.: A hierarchical multiscale cohesive zone model and simulation of dynamic fracture in metals. Eng. Fract. Mech. **163**, 327–347 (2016)
32. Liu, L., Li, S.: A finite temperature multiscale interphase zone model and simulations of fracture. J. Eng. Mater. Technol. **134**(3), 031014 (2012)
33. Tersoff, J.: Empirical interatomic potential for silicon with improved elastic properties. Phys. Rev. B **38**(14), 9902 (1988)

34. Li, S., Urata, S.: An atomistic-to-continuum molecular dynamics: theory, algorithm, and applications. Comput. Methods Appl. Mech. Eng. **306**, 452–478 (2016)
35. Parrinello, M., Rahman, A.: Polymorphic transitions in single crystals: a new molecular dynamics method. J. Appl. Phys. **52**(12), 7182–7190 (1981)
36. Chen, Y., Lee, J.D.: Multiscale modeling of polycrystalline silicon. Int. J. Eng. Sci. **42**(10), 987–1000 (2004)
37. Stemmer, S., Streiffer, S.K., Browning, N.D., Basceri, C., Kingon, A.I.: Grain boundaries in barium strontium titanate thin films: structure, chemistry and influence on electronic properties. Interface Sci. **8**(2–3), 209–221 (2000)

Critical Scaling and Prediction of Snap-Through Buckling

Tetsuo Yamaguchi, Hiroshi Ohtsubo and Yoshinori Sawae

Abstract We examined predictability of catastrophic events in snap-through buckling. In our study, we applied load at the center of an elastic arch and monitored the fundamental frequency by hammering. We found that the frequency goes to zero as the load approaches the critical value (load limit point), suggesting that the occurrence of the snap-through buckling accompanying an abrupt collapse of the structure can be predicted before it happens. We described the behavior with a simple theoretical model, derived scaling relations, and confirmed reasonable agreements with experimental results.

1 Introduction

There are various types of catastrophic phenomena in our daily lives, such as fractures, failures of machines, collapse of structures, earthquakes, and so on. All these phenomena have common features; before an event, energy is accumulated quasi-statically into the system and once it happens, the system rapidly release huge amount of energy leading to disastrous consequences [1].

Prediction of such catastrophic events, i.e., to estimate the timing and the magnitude before it happens, is one of the most important issues in engineering and science communities. In spite of many efforts made to find some clues of precursory phenomena in the field of structural health monitoring [2] or non-destructive testing [3], however, the prediction is thought to be difficult or even impossible at present.

On the other hand, there have been many discussions on bifurcation behavior in dynamical systems. In particular, the saddle node bifurcation exhibits striking behavior [4]; as the system approaches the bifurcation point, the stiffness of the system becomes smaller and vanishes as it reaches the bifurcation point. Based on the idea of critical softening, there have been some laboratory experiments conducted

T. Yamaguchi (✉) · H. Ohtsubo · Y. Sawae
Department of Mechanical Engineering, Kyushu University, 744 Motooka, Nishi-ku, Fukuoka 819-0395, Japan
e-mail: yamaguchi@mech.kyushu-u.ac.jp

P. van Meurs et al. (eds.), *Mathematical Analysis of Continuum Mechanics and Industrial Applications II*, Mathematics for Industry 30,
https://doi.org/10.1007/978-981-10-6283-4_5

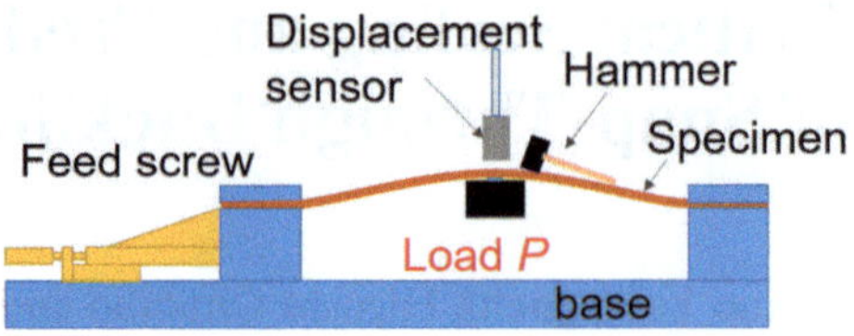

Fig. 1 Schematic of experimental setup

to predict catastrophe in simple settings, such as an Euler strut [4] and an arch-like structure [5, 6].

In this study, we re-examine the snap-through buckling experiments done by Wiebe et al. [5] where they observed the softening behavior as the system approaches the bifurcation point. We introduce the critical scaling to fit experimental data and develop a physical model to describe the behavior.

2 Experiment

Figure 1 shows our experimental setup (see also Fig. 4 for details). We used a flat elastic plate made of phosphor bronze ($L_0 = 680$ mm, $W_0 = 40$ mm, $H_0 = 1$ mm, Misumi, Japan) as a specimen. We firmly fixed the plate at both ends (clamped length $= 40$ mm for each end, thus the initial arch length becomes 600 mm) to satisfy the boundary conditions, i.e., the vertical displacement and its slope are set to zero at both ends. We then compressed the flat plate laterally by adjusting the position of the feed screw to cause the Euler buckling. The Euler buckling accompanies spontaneous symmetry breaking in terms of bending deformation: the plate buckles into a convex upward or downward shape. Here we selected only convex upward state as an arch structure, shown in the figure. We also applied the load P in this experiment. As we will see later, the arch structure causes snap-through buckling (jumps from convex upward to downward) at a critical load P_c. In order to study the dynamic response of the system before the snap-through buckling, we hammered the center of the arch as gentle as possible and measured the displacement at the center with a capacitance-type displacement sensor (VE-1020, Ono Sokki, Japan).

3 Results and Discussion

3.1 Buckling Experiment

Figure 2 shows typical examples of hammering experiments before the snap-through buckling. In both conditions (a) $P = 61.48$ g ($P \ll P_c$) and (b) 180.43 g ($P \approx P_c$), the arch oscillated for more than 5 s. From these displacement waveforms, we

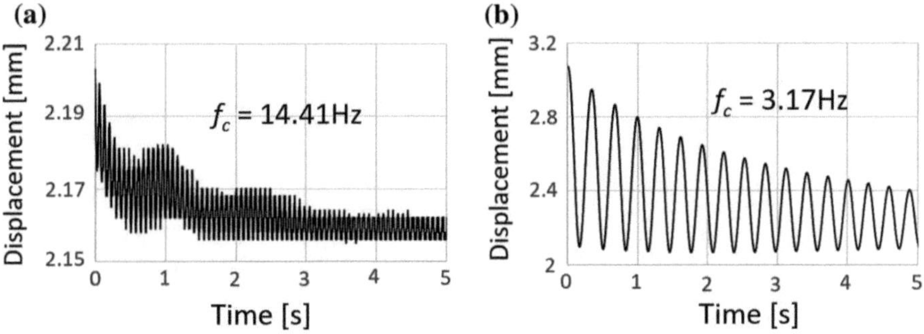

Fig. 2 Typical examples of time-displacement curves, **a** $P = 61.48$ g and **b** $P = 180.43$ g

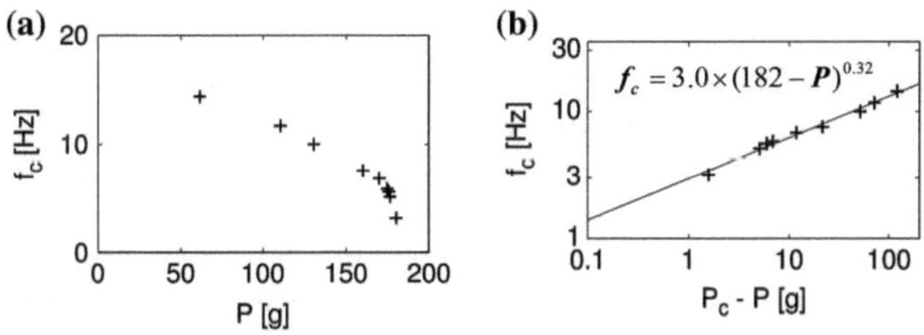

Fig. 3 **a** Relationship between P and f_c, and **b** between $P_c - P$ and f_c. P_c was estimated by the non-linear least square fitting of the equation $f_c = \beta(P_c - P)^\alpha$

obtained the frequency spectrum by applying the Fourier transform. We identified the lowest frequency mode f_c as the fundamental frequency, which was 14.41 and 3.17 Hz for (a) and (b), respectively.

In Fig. 3a, we plotted the relationship between the load P and the fundamental frequency f_c. As is shown clearly, the frequency decreases as we increase the load and interestingly, it approaches zero just before the system collapses ($P_c \approx 181.0$ g in the present experiment).

To characterize this asymptotic behavior, we applied the critical scaling. The relationship between P and f_c is described by the following formula.

$$f_c = \beta(P_c - P)^\alpha, \tag{1}$$

where α, β, and P_c are constants to be estimated. We applied the non-linear least square fitting and obtained the values $\alpha = 0.324$, $\beta = 2.95$, and $P_c = 182.0$ g. It is important to note that the critical load estimated by the fitting (182.0 g) is in good agreement with the actual value of collapse (181.0 g) with 1 g accuracy. This suggests that the state of the system can be monitored with the frequency response and also that the occurrence of collapse can be predicted just by extrapolating $P - f_c$ curve.

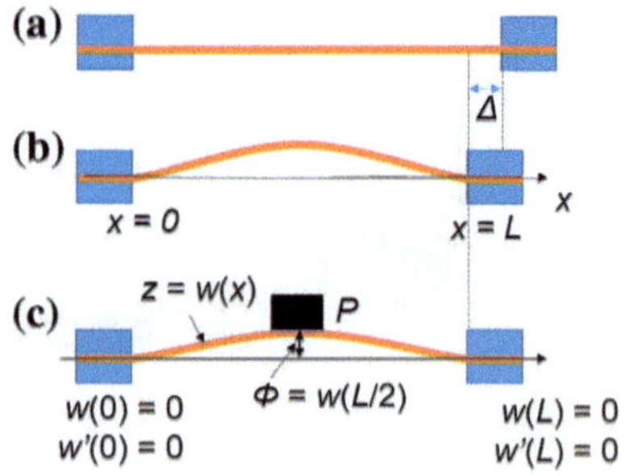

Fig. 4 Model description. **a** Initial state, **b** Euler buckled state, and **c** the state after load P is imposed

3.2 Theoretical Analysis

Let us then describe the critical behavior with a simple physical model. Figure 4 illustrates the geometry and the state of the system. In Fig. 4a, the system is in the initial state. We then apply lateral compression Δ to induce the Euler buckling, as shown in Fig. 4b. Furthermore, we impose load P at the center of the arch and resulting central deflection is $\Phi = w(L/2)$ (see Fig. 4c), where $w(x)$ is the vertical displacement of the plate at position x. As we mentioned before, we fixed the arch at both ends so that $w(0) = w'(0) = 0$ and $w(L) = w'(L) = 0$ (where $w'(x)$ represents the derivative of w with respect to x) are satisfied.

The total energy of the system is described by the following equation.

$$U_{total} = U_{bending} + U_{compression} + U_{gravity}$$
$$= \frac{EI}{2}\int_0^L \kappa^2(x')dx' + \frac{EAL}{2}\varepsilon^2 + Pw(L/2), \tag{2}$$

where E is the Young's modulus, $I = H_0^3 W_0/12$ is the second moment of area, and $A = H_0 W_0$ is the cross-section area of the arch. $\kappa(x)$, and ε are the curvature at position x and the lateral compressive strain, respectively, and are given as follows:

$$\kappa(x) = \frac{w''(x)}{\left[1 + \{w'(x)\}^2\right]^{3/2}} \approx w''(x), \tag{3}$$

$$\varepsilon = \frac{L + \Delta - \int_0^L \sqrt{1 + \{w'(x')\}^2}dx'}{L} \approx \frac{\Delta}{L} - \frac{1}{2L}\int_0^L \{w'(x')\}^2 dx', \tag{4}$$

where $w''(x)$ is the second-order derivative of w with respect to x. It is important to note that the approximations in Eqs. (3) and (4) hold for $w'(x) \ll 1$ (corresponding to the shallow arch [4]) and also that the lateral strain, derived by considering the contour length constraint [7], becomes constant (independent of x) due to the lateral force balance along x direction. Substituting Eqs. (3) and (4) into (2), we obtain the following equation.

$$U_{total} = \frac{EI}{2}\int_0^L \{w''(x')\}^2 dx' + \frac{EAL}{2}\left[\frac{\Delta}{L} - \frac{1}{2L}\int_0^L \{w'(x')\}^2 dx'\right]^2 + Pw(L/2). \tag{5}$$

Let us describe $w(x)$ as the linear combinations of the eigen functions for the homogeneous Sturm–Liouville problem [7];

$$w(x) = \sum_{i=0}^{\infty} w_i(x), \tag{6}$$

where $w_i(x) = A_i\{1 - cos(n_i x)\}$ for even i ($n_i = 2\pi/L, 4\pi/L, 6\pi/L, \ldots$ for $i = 0, 2, 4, \ldots$), and $w_i(x) = A_i\,[1 - cos(n_i x) - 2/(n_i L)\{n_i x - sin(n_i x)\}]$ for odd i ($n_i = 2.86\pi/L, 4.92\pi/L, 6.94\pi/L, \ldots$ for $i = 1, 3, 5, \ldots$). Note here that $w_i(x)$ satisfies the boundary conditions $w_i(0) = w_i'(0) = 0$ and $w_i(L) = w_i'(L) = 0$.

In principle, the problem can be solved by minimizing U_{total} in terms of A_i ($\partial U_{total}/\partial A_0 = \partial U_{total}/\partial A_1 = \partial U_{total}/\partial A_2 = \ldots = 0$). In order to simplify the discussion, however, we keep only the lowest mode ($i = 0$) and neglect other contributions from higher modes ($i > 0$). By using the relation $\Phi = w(L/2) = A_0 w_0(L/2) = 2A_0$, U_{total} can be rewritten as a function of the central deflection Φ:

$$U_{total} = \frac{\pi^2 E H_0 W_0}{32L^3}\Phi^4 + \frac{\pi E H_0 W_0}{12L^3}(\pi^3 H_0^2 - 3L\Delta)\Phi^2 + P\Phi + \frac{\Delta^2 E H_0 W_0}{2L}, \tag{7}$$

meaning that the system behavior can be described by what is called Landau theory in statistical physics [4] with the order parameter Φ. Thus, the state of the system is realized so that U_{total} becomes minimum in terms of Φ. By considering $\partial U_{total}/\partial\Phi = 0$, we obtain the following cubic equation.

$$\frac{\pi^2 E H_0 W_0}{8L^3}\Phi^3 + \frac{\pi E H_0 W_0}{6L^3}(\pi^3 H_0^2 - 3L\Delta)\Phi + P = 0. \tag{8}$$

When $P = 0$ (without load), we have two types of stable solutions depending on the magnitude of Δ:

- $\Phi = 0$ if $\Delta < \pi^3 H_0^2/(3L)$,
- $\Phi = \pm\sqrt{\frac{4}{3\pi}(3L\Delta - \pi^3 H_0^2)}$ otherwise,

where the first and the latter two solutions correspond to no buckled, convex upward, and downward states, respectively.

Next, let us consider the situation where $P > 0$ (with load). In this situation, we have two stable solutions and one unstable solution. Furthermore, at the critical point where the snap-through buckling occurs, the stable solutions should have a double root.

$$(\Phi - \Phi_c)^2(\Phi - \Phi_b) = 0, \tag{9}$$

where Φ_c and Φ_b are the displacement at the critical point and other one, respectively. By comparing the coefficient of each term, we obtain the following relations.

- Second order: $\Phi_b = -2\Phi_c$,
- First order: $\Phi_c = \sqrt{\frac{4}{3\pi}L(\Delta - \Delta_c)}$,
- Zeroth order: $P_c = \frac{\pi^2 E H_0 W_0}{4L^3}\Phi_c^3$.

From these relations, we obtain the formula for the stiffness $k = |\partial P/\partial \Phi|$ in the vicinity of the critical point.

$$k \approx \frac{3\pi^2 E H_0 W_0}{4L^3}|\Phi_c(\Phi - \Phi_c)| \approx \left\{\frac{3\pi^2 E H_0 W_0 \Phi_c}{2L^3}(P_c - P)\right\}^{\frac{1}{2}}, \tag{10}$$

and finally, we obtain the scaling relation between P and f_c.

$$f_c = 2\pi\sqrt{\frac{k}{P}} \sim (P_c - P)^{\frac{1}{4}}. \tag{11}$$

It is important to note that the theory reproduces experimentally obtained scaling form, and also that the exponent $\alpha = 0.25$ is in reasonable agreement with experimental values (0.324) though it is underestimated. This suggests that the scaling form can be used to predict (extrapolate) the critical point before a catastrophic buckling event happens.

4 Conclusion

We conducted buckling experiments of an elastic arch and investigated the possibility of high precision prediction for the snap-through buckling. We found that the occurrence of the snap-through buckling can be monitored and predicted with the lowest fundamental frequency of the arch. We also found that the frequency response obeys power-law scaling and the exponent is in agreement with the theoretical value. However, there remain some problems to be solved. For example, the discrepancy in the critical exponent between the theory and our experiment has to be corrected. Maybe dynamic modeling considering inertia effects improves the gap in between. This will be reported elsewhere in a separate paper.

Acknowledgements We thank T. Hatano for important suggestions. This work was supported by JSPS KAKENHI Grant Number JP16H06478 (Science of Slow Earthquakes).

References

1. Gilmore, R.: Catastrophe Theory for Scientists and Engineers. Dover Publications, New York (1993)
2. Balageas, D., Fritzen, C.-P., Guemes, A. (eds.): Structural Health Monitoring. ISTE, London (2006)
3. Halmshaw, R.: Non-Destructive Testing. Edward Arnold, London (1987)
4. Pippard, A.B.: Response and Stability: An Introduction to the Physical Theory. Cambridge University Press, Cambridge (1985)
5. Wiebe, R., Virgin, L., Stanciulescu, I., Spottswood, S.: On snap-through buckling. In: 52nd AIAA/ASME/ASCE/AHS/ASC Structures, Structural Dynamics and Materials Conference, vol. 2083 (2011)
6. Chandra, Y., Stanciulescu, I., Virgin, L.N., Eason, T.G., Spottswood, S.M.: A numerical investigation of snap-through in a shallow arch-like model. J. Sound Vib. **332**, 25322548 (2013)
7. Vangbo, M.: An analytical analysis of a compressed bistable buckled beam. Sens. Actuators A **69**, 212–216 (1998)

Part II
Shape Optimization

Second Derivatives of Cost Functions and H^1 Newton Method in Shape Optimization Problems

Hideyuki Azegami

Abstract We derive the second-order shape derivatives (shape Hessians) of cost functions for shape optimization problems of domains in which boundary value problems of partial differential equations are defined, and propose an H^1 Newton method to solve the problems using the shape Hessians. In this paper, we formulate an abstract shape optimization problem and show the computations of the first- and second-order shape derivatives of cost functions under the abstract framework. Then, using the shape gradients and Hessians, we propose an H^1 Newton method to solve the given problem. As an illustration, the shape Hessians of a mean compliance and a domain measure are derived and then used for a numerical example.

1 Introduction

Optimization problems with respect to shapes of domains in which boundary value problems of partial differential equations are defined are called shape optimization problems. Among them, a problem formulated using a domain variation by the Lagrange description as a design variable is classified as a domain variation type. In the previous paper [2], we showed an evaluation method of Fréchet derivatives (shape derivatives) of cost functions with respect to domain variation and a solution to the problems based on a gradient method (H^1 gradient method). In this paper, we present an evaluation method of the second-order shape derivatives (shape Hessians) of cost functions and a solution using a Newton method (H^1 Newton method).

Notation $W^{s,p}\left(\Omega_0; \mathbb{R}^d\right)$ will be used to represent the Sobolev space for the set of functions which are defined in Ω_0, have values in $\mathbb{R}^d$, and are $s \in [0, \infty]$ times differentiable and $p \in [1, \infty]$-th order Lebesgue integrable. Moreover, $L^p\left(\Omega_0; \mathbb{R}^d\right)$ and $H^s\left(\Omega_0; \mathbb{R}^d\right)$ are denoted by $W^{0,p}\left(\Omega_0; \mathbb{R}^d\right)$ and $W^{s,2}\left(\Omega_0; \mathbb{R}^d\right)$, respectively.

H. Azegami (✉)
Graduate School of Informatics, Nagoya University, A4-2(780) Furo-cho, Chikusa-ku, Nagoya 464-8601, Japan
e-mail: azegami@i.nagoya-u.ac.jp

P. van Meurs et al. (eds.), *Mathematical Analysis of Continuum Mechanics and Industrial Applications II*, Mathematics for Industry 30,
https://doi.org/10.1007/978-981-10-6283-4_6

2 Abstract Optimum Design Problem

In this paper, basic results will be shown using an abstract optimum design problem as follows. Let $\phi \in \mathscr{D} \subset X$ be a design variable from an admissible set $\mathscr{D}$ in a Hilbert space X. For $\phi \in \mathscr{D}$, a state variable u in a Hilbert space U is assumed to be uniquely determined as a solution of the following problem.

Problem 2.1 (*Abstract variational problem*) For $\phi \in \mathscr{D}$, let $a(\phi) : U \times U \to \mathbb{R}$ be a bounded and coercive bilinear form in U, and $l(\phi) = l(\phi)(\,\cdot\,) = \langle l(\phi), \,\cdot\,\rangle \in U'$ (dual space of U). Find $u \in U$ such that

$$a(\phi)(u, v) = l(\phi)(v) \quad \forall v \in U.$$

Problem 2.1 can be equivalently stated as follows. "Let $\tau(\phi) : U \to U'$ be an isomorphism by Lax–Milgram theorem (Theorem 18.F in [3]) when $a(\phi)(\,\cdot\,, \,\cdot\,)$ is a bounded and coercive bilinear form in U. Find $u \in U$ satisfying

$$s(\phi, u) = l(\phi) - \tau(\phi) u = 0_{U'}.\text{''} \tag{1}$$

We assume that the solution u of Problem 2.1 is an element of admissible set $\mathscr{S} \subset U$ in order to assure that $\phi + \varphi \in \mathscr{D}$, where φ is a variation of the design variable ϕ obtained by the gradient or the Newton method which will be shown later. Given the pair $(\phi, u) \in \mathscr{D} \times \mathscr{S}$, we consider the following design problem.

Problem 2.2 (*Abstract optimum design problem*) For $f_0, \ldots, f_m : \mathscr{D} \times \mathscr{S} \to \mathbb{R}$, find $(\phi^\star, u^\star) \in \mathscr{D} \times \mathscr{S}$ such that

$$f_0\left(\phi^\star, u^\star\right) = \min_{(\phi,u)\in\mathscr{D}\times\mathscr{S}} \left\{ f_0(\phi, u) \mid f_1(\phi, u) \le 0, \ldots, f_m(\phi, u) \le 0, \ \text{Problem 2.1} \right\}.$$

2.1 Gradient of Cost Function f_i with Respect to ϕ

In Problem 2.1, Problem 2.2 is assumed as an equality constraint. In this section, using the following problem, we show the computations of the Fréchet derivative and Hessian of cost function f_i with respect to an arbitrary variation $\varphi \in X$ (or $\varphi \in Y \subset X$ with a linear space $Y \supset \mathscr{D}$) of the design variable $\phi \in \mathscr{D}$ subject to the given equality constraint.

Problem 2.3 (*Abstract optimum design problem with equality constraint*) For $f_i : \mathscr{D} \times \mathscr{S} \to \mathbb{R}$ and $s(\phi, u)$ in (1), find $(\phi^\star, u^\star)$ such that

$$f_i\left(\phi^\star, u^\star\right) = \min_{(\phi,u)\in\mathscr{D}\times\mathscr{S}} \left\{ f_i(\phi, u) \mid s(\phi, u) = 0_{U'} \right\}.$$

In order to show necessary and sufficient conditions for a local minimizer of Problem 2.3, we define the Lagrange function with respect to Problem 2.3 as

$$\mathscr{L}_i(\phi, u, v_i) = f_i(\phi, u) + \langle s(\phi, u), v_i \rangle = f_i(\phi, u) + \mathscr{L}_S(\phi, u, v_i), \tag{2}$$

where $\mathscr{L}_S(\phi, u, v_i)$ denotes the Lagrange function with respect to Problem 2.1, and u and v_i are the variables in $\mathscr{S} \subset U$ corresponding to the solution of Problem 2.1 and the Lagrange multiplier with respect to the equality constraint of Problem 2.1 for f_i, respectively. In the definition of Lagrange function of (2), notice that u is not necessary to be the solution of Problem 2.1. $\mathscr{S}$ and U will be used as the admissible set and the set for test function of u and v_i in Problem 2.1 and the adjoint problem shown later in (13), respectively. With respect to an arbitrary variation $(\varphi, \hat{u}, \hat{v}_i) \in X \times U^2$ of $(\phi, u, v_i) \in \mathscr{D} \times \mathscr{S}^2$, the Fréchet derivative of $\mathscr{L}_i$ can be written as

$$\begin{aligned}
&\mathscr{L}_i'(\phi, u, v_i)[\varphi, \hat{u}, \hat{v}_i] = \mathscr{L}_{i\phi}(\phi, u, v_i)[\varphi] + \mathscr{L}_{iu}(\phi, u, v_i)[\hat{u}] + \mathscr{L}_{iv_i}(\phi, u, v_i)[\hat{v}_i] \\
&= f_i'(\phi, u)[\varphi, \hat{u}] + \langle s'(\phi, u)[\varphi, \hat{u}], v_i \rangle + \langle s(\phi, u), \hat{v}_i \rangle \\
&= f_{i\phi}(\phi, u)[\varphi] + f_{iu}(\phi, u)[\hat{u}] + \langle s_\phi(\phi, u)[\varphi] + s_u(\phi, u)[\hat{u}], v_i \rangle + \langle s(\phi, u), \hat{v}_i \rangle \\
&= (\langle f_{i\phi}(\phi, u), \varphi \rangle + \langle s_\phi(\phi, u)[\varphi], v_i \rangle) + \langle f_{iu}(\phi, u) - \tau^*(\phi) v_i, \hat{u} \rangle + \langle s(\phi, u), \hat{v}_i \rangle \\
&= \langle g_i(\phi, u, v_i), \varphi \rangle + \mathscr{L}_{iu}(\phi, u, v_i)[\hat{u}] + \mathscr{L}_S(\phi, u, \hat{v}_i).
\end{aligned} \tag{3}$$

Here, $\tau^*(\phi) : U \to U'$ is the adjoint operator of $\tau(\phi)$. Using the notations in (3), we have the following result.

Theorem 2.1 (The first-order necessary condition of a local minimizer) *Let $f_i \in C^1(\mathscr{D} \times \mathscr{S}; \mathbb{R})$ and $s \in C^1(\mathscr{D} \times \mathscr{S}; U')$. If (ϕ, u) is a local minimizer of Problem 2.3, there exists a $v_i \in U$ which satisfies*

$$\langle g_i(\phi, u, v_i), \varphi \rangle + \mathscr{L}_{iu}(\phi, u, v_i)[\hat{u}] = 0 \quad \forall (\varphi, \hat{u}) \in X \times U, \tag{4}$$

$$\mathscr{L}_S(\phi, u, \hat{v}_i) = 0 \quad \forall \hat{v}_i \in U. \tag{5}$$

Proof From the fact that $s \in C^1(\mathscr{D} \times \mathscr{S}; U')$ and there is a unique solution u which satisfies $s(\phi, u) = 0_{U'}$, s satisfies the following assumptions for the implicit function theorem in a neighborhood $B_X \times B_U \subset X \times U$ of $(\phi, u) \in \mathscr{D} \times \mathscr{S}$:

1. $s(\phi, u) = 0_{U'}$,
2. $s \in C^0(B_X \times B_U; U')$,
3. $s(\phi, \cdot) \in C^1(B_U; U')$ with respect to an arbitrary $y = (\phi, w) \in B_X \times B_U$ and $s_u(\phi, u) = -\tau : U \to U'$ is continuous at (ϕ, u),
4. $(s_u(\phi, u))^{-1} = -\tau^{-1} : U' \to U$ is bounded and linear.

From the implicit function theorem, there exist some neighborhood $\hat{B}_X \times \hat{B}_U \subset B_X \times B_U$ and continuous mapping $\upsilon : \hat{B}_X \to \hat{B}_U$, and $s(\phi, u) = 0_{U'}$ can be written as

$$u = \upsilon(\phi). \tag{6}$$

Hence, we can define $y(\phi) = (\phi, \upsilon(\phi)) \in C^1(\mathscr{D}; X \times U)$ and write $f_i(\phi, \upsilon(\phi)) = f_i(y(\phi))$ by $\tilde{f}_i(\phi)$. Since $f_i \in C^1(\mathscr{D} \times \mathscr{S}; \mathbb{R})$, then ϕ being a local minimizer implies that

$$\tilde{f}_i'(\phi)[\varphi] = y'^*(\phi) \circ g_i(\phi, \upsilon(\phi))[\varphi] = 0 \quad \forall \varphi \in X. \tag{7}$$

Here,

$$g_i(\phi, \upsilon(\phi)) = f_i'(\phi, \upsilon(\phi)) \in \mathfrak{L}(X; X' \times U') = \mathfrak{L}(X; \mathfrak{L}(X \times U; \mathbb{R})),$$
$$y'(\phi) \in \mathfrak{L}(X; X \times U), \quad y'^*(\phi) \in \mathfrak{L}(X' \times U'; X').$$

In this paper, $\mathfrak{L}(X; U)$ denotes the set of all bounded linear operators from X to U, and $\circ$ is the composition operator. We rewrite (7) as below. Firstly, let's write the admissible set of (ϕ, u) with respect to the equality constraint as

$$S = \{(\phi, u) \in \mathscr{D} \times \mathscr{S} \mid s(\phi, u) = 0_{U'}\}. \tag{8}$$

For $y(\phi) = (\phi, u) \in S$, we denote the kernel of $s'(\phi, u) \in \mathfrak{L}(X \times U; U')$ by

$$T_S(\phi, u) = \{(\varphi, \hat{\upsilon}) \in X \times U \mid s'(\phi, u)[\varphi, \hat{\upsilon}] = 0_{U'}\} \tag{9}$$

and the space orthogonal to $T_S(\phi, u)$ as

$$T_S'(\phi, u) = \{(\psi, w) \in X' \times U' \mid \langle(\varphi, \hat{\upsilon}), (\psi, w)\rangle = 0 \, \forall (\varphi, \hat{\upsilon}) \in T_S(\phi, u)\}.$$

Moreover, the relationship between $T_S(\phi, u)$ and the Fréchet derivative $y'(\phi)[\varphi]$ of $y(\phi) \in S$ with respect to an arbitrary variation $\varphi \in X$ can be obtained in the following way. If we take the Fréchet derivative on both sides of $s(\phi, u) = 0_{U'}$ with respect to $\varphi \in X$, then we get

$$s'(\phi, u) \circ y'(\phi)[\varphi] = 0_{U'} \quad \forall \varphi \in X. \tag{10}$$

This relationship shows that the image space $\operatorname{Im} y'(\phi)$ of $y'(\phi)$ is actually the kernel space $\operatorname{Ker} s'(\phi, u)$ of $s'(\phi, u)$. In other words, the following relationship is established:

$$T_S(\phi, u) = \operatorname{Im} y'(\phi). \tag{11}$$

We use the relationship above to rewrite (7). When ϕ is a local minimizer, $g_i(\phi, \upsilon(\phi))$ needs to be orthogonal to an arbitrary $(\varphi, v_i) \in T_S(\phi, u)$. Hence,

$$g_i(\phi, \upsilon(\phi)) \in T_S'(\phi, u). \tag{12}$$

Now, using the relationship between the orthogonal complement space of the image space and the kernel space, we get that $T_S'(\phi, u) = \operatorname{Im} s'^*(\phi, u)$ where $s'^*(\phi, u) \in$

$\mathfrak{L}\left(U; X' \times U'\right)$ is the adjoint of $s'(\phi, u)$. Therefore, (12) means that we can find an element v_i of U such that

$$f_i'(\phi, u)\left[\varphi, \hat{u}\right] + \left\langle s'(\phi, u)\left[\varphi, \hat{u}\right], v_i\right\rangle = 0 \quad \forall \left(\varphi, \hat{u}\right) \in X \times U.$$

This established Eq. (4). Moreover, (5) holds if u is the solution of (1). (QED)

From Theorem 2.1, g_i can be evaluated in the following way. Let u be determined as satisfying (5). This means u is the solution of Problem 2.1. Moreover, let v_i be determined through the equation

$$\mathscr{L}_{iu}(\phi, u, v_i)\left[\hat{u}\right] = \left\langle f_{iu}(\phi, u) - \tau^*(\phi) v_i, \hat{u}\right\rangle = 0 \quad \forall \hat{u} \in U. \tag{13}$$

This problem to determine v_i by (13) is called the adjoint problem of Problem 2.1 with respect to f_i. When we use the solutions u and v_i, $g_i \in X'$ is obtained through

$$\langle g_i(\phi, u, v_i), \varphi\rangle = \left\langle f_{i\phi}(\phi, u), \varphi\right\rangle + \left\langle s_\phi(\phi, u)[\varphi], v_i\right\rangle \quad \forall \varphi \in X. \tag{14}$$

2.2 Hessian of Cost Function f_i with Respect to Variation of ϕ

Furthermore, when f_i and s are the second-order Fréchet differentiable, then we may calculate the second-order partial Fréchet derivative of $\mathscr{L}_i$ with respect to arbitrary variations $\left(\varphi_1, \hat{\upsilon}_1\right), \left(\varphi_2, \hat{\upsilon}_2\right) \in T_S(\phi, u)$ of $(\phi, u) \in S$ as

$$\begin{aligned}&\mathscr{L}_{i(\phi,u),(\phi,u)}(\phi, u, v_i)\left[\left(\varphi_1, \hat{\upsilon}_1\right), \left(\varphi_2, \hat{\upsilon}_2\right)\right]\\ &\quad = f_i''(\phi, u)\left[\left(\varphi_1, \hat{\upsilon}_1\right), \left(\varphi_2, \hat{\upsilon}_2\right)\right] + \left\langle s''(\phi, u)\left[\left(\varphi_1, \hat{\upsilon}_1\right), \left(\varphi_2, \hat{\upsilon}_2\right)\right], v_i\right\rangle.\end{aligned} \tag{15}$$

Using it, we have the following result.

Theorem 2.2 (The second-order necessary condition of a local minimizer) *Let f_i and s be elements of $C^2(\mathscr{D} \times \mathscr{S}; \mathbb{R})$ and $C^2\left(\mathscr{D} \times \mathscr{S}; U'\right)$, respectively. If (ϕ, u) is a local minimizer of Problem 2.3, the following holds:*

$$\mathscr{L}_{i(\phi,u),(\phi,u)}(\phi, u, v_i)\left[\left(\varphi, \hat{\upsilon}\right), \left(\varphi, \hat{\upsilon}\right)\right] \geq 0 \quad \forall \left(\varphi, \hat{\upsilon}\right) \in T_S(\phi, u). \tag{16}$$

Proof In the proof of Theorem 2.1, the assumption for the implicit function theorem is replaced by $s(\phi, \cdot) \in C^2\left(B_U; U'\right)$, and then using $\upsilon(\phi)$ in (6), $y(\phi) = (\phi, \upsilon(\phi)) \in C^2(\mathscr{D}; X \times U)$ is determined. From (10), we have

$$s''(\phi, u)\left[y'(\phi)[\varphi], y'(\phi)[\varphi]\right] = 0_{U'} \tag{17}$$

with respect to $y'(\phi)[\varphi] \in T_S(\phi, u)$. Hence, if (ϕ, u) is a local minimizer of Problem 2.3,

$$\mathscr{L}_{i(\phi,u),(\phi,u)}(\phi,u,v_i)\left[y'(\phi)[\varphi],y'(\phi)[\varphi]\right]=\tilde{f}_i''(\phi)[\varphi,\varphi]\geq 0 \tag{18}$$

holds with respect to $y'(\phi)[\varphi]\in T_S(\phi,u)$. (QED)

Since the left-hand side of (18) is the Hessian of f_i with respect to an arbitrary variation $\varphi\in X$ of ϕ, we write it as $h_i(\phi,u,v_i)[\varphi,\varphi]$, and obtain the following result.

Theorem 2.3 (The second-order sufficient condition for a local minimizer) *Under the assumptions of Theorem 2.2, if (4) and (5) are satisfied at $(\phi,u,v_i)\in\mathscr{D}\times\mathscr{S}^2$ and (16) holds, then (ϕ,u) is a local minimizer of Problem 2.3.*

Proof When $(\phi,u,v_i)\in\mathscr{D}\times\mathscr{S}^2$ is a stationary point of $\mathscr{L}_i$ in S, with respect to an arbitrary point $y(\phi+\varphi)=y(\phi)+z(\varphi)$ in a neighborhood $B\subset S$ of $y(\phi)=(\phi,u)$, there exists a $\theta\in(0,1)$ satisfying

$$\tilde{f}_i(\phi+\varphi)-\tilde{f}_i(\phi)=\frac{1}{2}\mathscr{L}_i''(\phi+\theta\varphi,u,v_i)[z(\varphi),z(\varphi)]\quad\forall y(\phi)+z(\varphi)\in B.$$

From the assumption, since the right-hand side is greater than or equal to 0, $\tilde{f}_i(\phi)\leq\tilde{f}_i(\phi+\varphi)$ holds. (QED)

In view of Theorems 2.2 and 2.3, h_i is calculated as

$$\begin{aligned}h_i(\phi,u,v_i)[\varphi_1,\varphi_2]&=\left(\mathscr{L}_{i\phi}(\phi,u,v_i)[\varphi_1]+\mathscr{L}_{iu}(\phi,u,v_i)\left[\hat{\upsilon}_1\right]\right)_\phi[\varphi_2]\\&\quad+\left(\mathscr{L}_{i\phi}(\phi,u,v_i)[\varphi_1]+\mathscr{L}_{iu}(\phi,u,v_i)\left[\hat{\upsilon}_1\right]\right)_u\left[\hat{\upsilon}_2\right]\\&=\mathscr{L}_{i\phi\phi}(\phi,u,v_i)[\varphi_1,\varphi_2]+\mathscr{L}_{iu\phi}(\phi,u,v_i)\left[\hat{\upsilon}_1,\varphi_2\right]\\&\quad+\mathscr{L}_{i\phi u}(\phi,u,v_i)\left[\varphi_1,\hat{\upsilon}_2\right]+\mathscr{L}_{iuu}(\phi,u,v_i)\left[\hat{\upsilon}_1,\hat{\upsilon}_2\right],\end{aligned} \tag{19}$$

where, in order that $\left(\varphi_j,\hat{\upsilon}_j\right)\in T_S(\phi,u)$ for $j\in\{1,2\}$, $\hat{\upsilon}_j=\upsilon'(\phi)\left[\varphi_j\right]$ has to be determined using the equation

$$\mathscr{L}_{\mathrm{S}(\phi,u)}(\phi,u,v)\left[\varphi_j,\hat{\upsilon}_j\right]=0\quad\forall\varphi_j\in X. \tag{20}$$

3 Shape Optimization Problem of Linear Elastic Body

Based on the results above, we illustrate how to find a solution to a shape optimization problem using Hessians. In particular, we consider a shape optimization problem of a d-dimensional linear elastic body ($d\in\{2,3\}$). In such a case of shape optimization problem of domain variation type, we set a domain variation (displacement) $\boldsymbol{\phi}$ from an initial domain Ω_0 as a design variable and denote the varied domain with $\Omega(\boldsymbol{\phi})$. Similarly, we denote $(\,\cdot\,)(\boldsymbol{\phi})$ as the varied. The function spaces for $\boldsymbol{\phi}$ are defined as $X=\{\boldsymbol{\phi}\in H^1\left(\mathbb{R}^d;\mathbb{R}^d\right)\mid\boldsymbol{\phi}=\mathbf{0}_{\mathbb{R}^d}\text{ on }\bar{\Omega}_{\mathrm{C0}}\}$, where $\bar{\Omega}_{\mathrm{C0}}$ expresses a part of boundary or domain to fix while the domain varies, $Y=X\cap W^{1,\infty}\left(\mathbb{R}^d;\mathbb{R}^d\right)$ and

$\mathscr{D} = \{\boldsymbol{\phi} \in Y \mid \Omega_0 \to \Omega(\boldsymbol{\phi}) \text{ is a bijection}\}$. Let $\mathbf{u} : \Omega(\boldsymbol{\phi}) \to \mathbb{R}^d$ be an elastic displacement, $\mathbf{E}(\mathbf{u}) = \{\boldsymbol{\nabla}\mathbf{u}^{\mathrm{T}} + (\boldsymbol{\nabla}\mathbf{u}^{\mathrm{T}})^{\mathrm{T}}\}/2$ be a linear strain, $\mathbf{S}(\mathbf{u}) = \mathbf{C}(\boldsymbol{\phi})\mathbf{E}(\mathbf{u})$ be a stress, and $\mathbf{C}(\boldsymbol{\phi}) : \Omega(\boldsymbol{\phi}) \to \mathbb{R}^{d\times d\times d\times d}$ denote a stiffness. We use $\boldsymbol{\nu}$ as the normal.

Problem 3.1 (*Linear elastic problem*) For $\boldsymbol{\phi} \in \mathscr{D}, \mathbf{b}(\boldsymbol{\phi}), \mathbf{p}_{\mathrm{N}}(\boldsymbol{\phi}), \mathbf{u}_{\mathrm{D}}(\boldsymbol{\phi}) : \mathbb{R}^d \to \mathbb{R}^d$ and $\mathbf{C}(\boldsymbol{\phi}) : \mathbb{R}^d \to \mathbb{R}^{d\times d\times d\times d}$, find $\mathbf{u} : \Omega(\boldsymbol{\phi}) \to \mathbb{R}^d$ such that

$$-\boldsymbol{\nabla}^{\mathrm{T}}\mathbf{S}(\boldsymbol{\phi}, \mathbf{u}) = \mathbf{b}^{\mathrm{T}}(\boldsymbol{\phi}) \quad \text{in } \Omega(\boldsymbol{\phi}), \quad \mathbf{S}(\boldsymbol{\phi}, \mathbf{u})\boldsymbol{\nu} = \mathbf{p}_{\mathrm{N}}(\boldsymbol{\phi}) \quad \text{on } \Gamma_p(\boldsymbol{\phi}),$$
$$\mathbf{S}(\boldsymbol{\phi}, \mathbf{u})\boldsymbol{\nu} = \mathbf{0}_{\mathbb{R}^d} \quad \text{on } \Gamma_{\mathrm{N}}(\boldsymbol{\phi}) \setminus \bar{\Gamma}_p(\boldsymbol{\phi}), \quad \mathbf{u} = \mathbf{u}_{\mathrm{D}}(\boldsymbol{\phi}) \quad \text{on } \Gamma_{\mathrm{D}}(\boldsymbol{\phi}).$$

When the given functions are given appropriately, $\mathbf{u} - \mathbf{u}_{\mathrm{D}}$ is found uniquely in $\mathscr{S} = U \cap W^{1,\infty}(\Omega(\boldsymbol{\phi}); \mathbb{R}^d)$ with $U = \{\mathbf{u} \in H^1(\Omega(\boldsymbol{\phi}); \mathbb{R}^d) \mid \mathbf{u} = \mathbf{0}_{\mathbb{R}^d} \text{ on } \Gamma_{\mathrm{D}}(\boldsymbol{\phi})\}$. Using $\boldsymbol{\phi}$ and $\mathbf{u}$, we define an objective cost function f_0, which is called a mean compliance, and a constraint cost function f_1 for domain measure as specified below, and set a shape optimization problem in Problem 3.2.

$$f_0(\boldsymbol{\phi}, \mathbf{u}) = \int_{\Omega(\boldsymbol{\phi})} \mathbf{b} \cdot \mathbf{u} \, \mathrm{d}x + \int_{\Gamma_p(\boldsymbol{\phi})} \mathbf{p}_{\mathrm{N}} \cdot \mathbf{u} \, \mathrm{d}\gamma - \int_{\Gamma_{\mathrm{D}}(\boldsymbol{\phi})} \mathbf{u}_{\mathrm{D}} \cdot (\mathbf{S}(\mathbf{u})\boldsymbol{\nu}) \, \mathrm{d}\gamma, \tag{21}$$

$$f_1(\boldsymbol{\phi}) = \int_{\Omega(\boldsymbol{\phi})} \mathrm{d}x - c_1. \tag{22}$$

Problem 3.2 (*Mean compliance minimization*) Find $\Omega(\boldsymbol{\phi})$ satisfying

$$\min_{(\boldsymbol{\phi}, \mathbf{u}-\mathbf{u}_{\mathrm{D}}) \in \mathscr{D}\times\mathscr{S}} \{ f_0(\boldsymbol{\phi}, \mathbf{u}) \mid f_1(\boldsymbol{\phi}) \leq 0, \text{ Problem 3.1} \}.$$

3.1 Shape Gradients of Cost Functions

The shape gradient $\mathbf{g}_0$ of f_0 can be obtained using the relations of (13) and (14), where the Lagrange function $\mathscr{L}_{\mathrm{S}}$ for Problem 3.1 is given as

$$\mathscr{L}_{\mathrm{S}}(\boldsymbol{\phi}, \mathbf{u}, \mathbf{v}_0) = l(\phi)(\mathbf{v}_0) - a(\phi)(\tilde{\mathbf{u}}, \mathbf{v}_0),$$

where $\tilde{\mathbf{u}} = \mathbf{u} - \mathbf{u}_{\mathrm{D}}$, $\mathbf{v}_0 \in U$ and

$$a(\phi)(\tilde{\mathbf{u}}, \mathbf{v}_0) = \int_{\Omega(\boldsymbol{\phi})} \mathbf{S}(\tilde{\mathbf{u}}) \cdot \mathbf{E}(\mathbf{v}_0) \, \mathrm{d}x,$$

$$l(\phi)(\mathbf{v}_0) = \int_{\Omega(\boldsymbol{\phi})} \mathbf{b} \cdot \mathbf{v}_0 \mathrm{d}x + \int_{\Gamma_p(\boldsymbol{\phi})} \mathbf{p}_{\mathrm{N}} \cdot \mathbf{v}_0 \, \mathrm{d}\gamma - a(\phi)(\mathbf{u}_{\mathrm{D}}, \mathbf{v}_0).$$

From the adjoint problem corresponding to (13), the self-adjoint condition $\mathbf{v}_0 = \mathbf{u}$ is obtained. Here, we assume $\mathbf{b}(\boldsymbol{\phi})$, $\mathbf{p}_{\mathrm{N}}(\boldsymbol{\phi})$, $\mathbf{u}_{\mathrm{D}}(\boldsymbol{\phi})$, and $\mathbf{C}(\boldsymbol{\phi})$ vary with domain variation. Then, using Propositions A.1 and A.2 in [2] (Propositions 9.3.4 and 9.3.7

in [1]), we have

$$\mathscr{L}_{0\boldsymbol{\phi}}(\boldsymbol{\phi}, \mathbf{u}, \mathbf{v}_0)[\boldsymbol{\varphi}] = \langle \mathbf{g}_0, \boldsymbol{\varphi} \rangle = \int_{\Omega(\boldsymbol{\phi})} \left(\mathbf{G}_{\Omega 0} \cdot \boldsymbol{\nabla} \boldsymbol{\varphi}^{\mathrm{T}} + g_{\Omega 0} \boldsymbol{\nabla} \cdot \boldsymbol{\varphi} \right) \mathrm{d}x + \int_{\Gamma_p(\boldsymbol{\phi})} \mathbf{g}_{p0} \cdot \boldsymbol{\varphi} \, \mathrm{d}\gamma + \int_{\partial \Gamma_p(\boldsymbol{\phi}) \cup \Theta(\boldsymbol{\phi})} \mathbf{g}_{\partial p0} \cdot \boldsymbol{\varphi} \, \mathrm{d}\varsigma \quad \forall \boldsymbol{\varphi} \in Y,$$

where

$$\mathbf{G}_{\Omega 0} = 2\mathbf{S}(\mathbf{u}) \left(\boldsymbol{\nabla} \mathbf{u}^{\mathrm{T}} \right)^{\mathrm{T}}, \quad g_{\Omega 0} = -\mathbf{S}(\mathbf{u}) \cdot \mathbf{E}(\mathbf{u}) + 2\mathbf{b} \cdot \mathbf{u},$$
$$\mathbf{g}_{p0} = 2\kappa (\mathbf{p}_{\mathrm{N}} \cdot \mathbf{u}) \boldsymbol{\nu}, \quad \mathbf{g}_{\partial p0} = 2(\mathbf{p}_{\mathrm{N}} \cdot \mathbf{u}) \boldsymbol{\tau}.$$

$\Theta(\boldsymbol{\phi})$ is the set of corner points in $\Gamma_p(\boldsymbol{\phi})$ for $d = 2$ and the vertices and edges for $d = 3$, $\boldsymbol{\tau}$ denotes the tangent, and $\kappa = \boldsymbol{\nabla} \cdot \boldsymbol{\nu}$. We can confirm that $\mathbf{g}_0 \in X'$ when $\mathbf{u} \in \mathscr{S}$ (Theorem 9.7.2 in [1]).

On the other hand, the shape derivative of f_1 is obtained as

$$f_1'(\boldsymbol{\phi})[\boldsymbol{\varphi}] = \langle \mathbf{g}_1, \boldsymbol{\varphi} \rangle = \int_{\Omega(\boldsymbol{\phi})} \boldsymbol{\nabla} \cdot \boldsymbol{\varphi} \, \mathrm{d}x \quad \forall \boldsymbol{\varphi} \in Y.$$

3.2 Shape Hessians of Cost Functions

The shape Hessian h_0 of f_0 is obtained using (19) and (20). In order to get a Hessian form, we assume thats $\mathbf{b} = \mathbf{0}_{\mathbb{R}^d}$ and Γ_{p0} is a fixed boundary, i.e., $\Gamma_{p0} \subset \bar{\Omega}_{\mathrm{C}0}$. In this case, (19) becomes

$$h_0(\boldsymbol{\phi}, \mathbf{u}, \mathbf{v}_0)[\boldsymbol{\varphi}_1, \boldsymbol{\varphi}_2] = \mathscr{L}_{0\boldsymbol{\phi}\boldsymbol{\phi}}(\boldsymbol{\phi}, \mathbf{u}, \mathbf{v}_0)[\boldsymbol{\varphi}_1, \boldsymbol{\varphi}_2] + \mathscr{L}_{0\boldsymbol{\phi}'\mathbf{u}}(\boldsymbol{\phi}, \mathbf{u}, \mathbf{v}_0)[\boldsymbol{\varphi}_1, \hat{\boldsymbol{v}}_2] + \mathscr{L}_{0\boldsymbol{\phi}'\mathbf{u}}(\boldsymbol{\phi}, \mathbf{u}, \mathbf{v}_0)[\boldsymbol{\varphi}_2, \hat{\boldsymbol{v}}_1] + \mathscr{L}_{0\mathbf{u}\mathbf{u}}(\boldsymbol{\phi}, \mathbf{u}, \mathbf{v}_0)[\hat{\boldsymbol{v}}_1, \hat{\boldsymbol{v}}_2]. \tag{23}$$

Corresponding to (20), using Proposition A.1 and A.2 in [2], for $j \in \{1, 2\}$, we get

$$\begin{aligned}
\mathscr{L}_{\mathrm{S}(\boldsymbol{\phi},\mathbf{u})}(\boldsymbol{\phi}, \mathbf{u}, \mathbf{v}_0)[\boldsymbol{\varphi}_j, \hat{\boldsymbol{v}}_j] &= \int_{\Omega(\boldsymbol{\phi})} \Big\{ \mathbf{S}(\mathbf{u}) \cdot \left(\boldsymbol{\nabla} \boldsymbol{\varphi}_j^{\mathrm{T}} \boldsymbol{\nabla} \mathbf{v}_0^{\mathrm{T}} \right)^{\mathrm{s}} + \mathbf{S}(\mathbf{v}_0) \cdot \left(\boldsymbol{\nabla} \boldsymbol{\varphi}_j^{\mathrm{T}} \boldsymbol{\nabla} \mathbf{u}^{\mathrm{T}} \right)^{\mathrm{s}} \\
&\quad - (\mathbf{S}(\mathbf{u}) \cdot \mathbf{E}(\mathbf{v}_0)) \boldsymbol{\nabla} \cdot \boldsymbol{\varphi}_j - \mathbf{S}(\hat{\boldsymbol{v}}_j) \cdot \mathbf{E}(\mathbf{v}_0) \Big\} \mathrm{d}x \\
&= \int_{\Omega(\boldsymbol{\phi})} \left[\left\{ \left(\boldsymbol{\nabla} \boldsymbol{\varphi}_j^{\mathrm{T}} \right)^{\mathrm{T}} \mathbf{S}(\mathbf{u}) + \mathbf{C} \left(\boldsymbol{\nabla} \boldsymbol{\varphi}_j^{\mathrm{T}} \boldsymbol{\nabla} \mathbf{u}^{\mathrm{T}} \right)^{\mathrm{s}} - \mathbf{S}(\mathbf{u}) \boldsymbol{\nabla} \cdot \boldsymbol{\varphi}_j - \mathbf{S}(\hat{\boldsymbol{v}}_j) \right\} \left(\boldsymbol{\nabla} \mathbf{v}_0^{\mathrm{T}} \right)^{\mathrm{T}} \right] \cdot \mathbf{I} \, \mathrm{d}x \\
&= 0.
\end{aligned} \tag{24}$$

Here, $(\cdot)^{\mathrm{s}} = \{(\cdot) + (\cdot)^{\mathrm{T}}\}/2$, the unit matrix $\mathbf{I} \in \mathbb{R}^{d \times d}$ and the Dirichlet boundary conditions for $\mathbf{v}_0$ and $\hat{\boldsymbol{v}}_j = \boldsymbol{v}'(\boldsymbol{\phi})[\boldsymbol{\varphi}_j]$ were used. From (24), the identity

$$\mathbf{S}\left(\hat{\boldsymbol{v}}_j\right)\left(\nabla \mathbf{v}_0^{\mathrm{T}}\right)^{\mathrm{T}} = \left\{\left(\nabla \boldsymbol{\varphi}_j^{\mathrm{T}}\right)^{\mathrm{T}} \mathbf{S}(\mathbf{u}) + \mathbf{C}\left(\nabla \boldsymbol{\varphi}_j^{\mathrm{T}} \nabla \mathbf{u}^{\mathrm{T}}\right)^{\mathrm{s}} - \nabla \cdot \boldsymbol{\varphi}_j \mathbf{S}(\mathbf{u})\right\}\left(\nabla \mathbf{v}_0^{\mathrm{T}}\right)^{\mathrm{T}} \tag{25}$$

is obtained. Moreover, since (24) can be written as

$$\begin{aligned}
&\mathscr{L}_{\mathrm{S}(\boldsymbol{\phi},\mathbf{u})}\left(\boldsymbol{\phi}, \mathbf{u}, \mathbf{v}_0\right)\left[\boldsymbol{\varphi}_j, \hat{\boldsymbol{v}}_j\right] \\
&= \int_{\Omega(\boldsymbol{\phi})}\left[\nabla \mathbf{v}_0^{\mathrm{T}} \mathbf{S}(\mathbf{u}) \nabla \boldsymbol{\varphi}_j^{\mathrm{T}} + \mathbf{S}(\mathbf{v}_0)\left\{\left(\nabla \mathbf{u}^{\mathrm{T}}\right)^{\mathrm{T}}\left(\left(\nabla \boldsymbol{\varphi}_j^{\mathrm{T}}\right)^{\mathrm{T}} - \nabla \cdot \boldsymbol{\varphi}_j\right) - \left(\nabla \hat{\boldsymbol{v}}_j^{\mathrm{T}}\right)^{\mathrm{T}}\right\}\right] \cdot \mathbf{I}\, \mathrm{d}x \\
&= 0,
\end{aligned}$$

we have another relation as

$$\mathbf{S}(\mathbf{v}_0)\left(\nabla \hat{\boldsymbol{v}}_j^{\mathrm{T}}\right)^{\mathrm{T}} = \nabla \mathbf{v}_0^{\mathrm{T}} \mathbf{S}(\mathbf{u}) \nabla \boldsymbol{\varphi}_j^{\mathrm{T}} + \mathbf{S}(\mathbf{v}_0)\left(\nabla \mathbf{u}^{\mathrm{T}}\right)^{\mathrm{T}}\left\{\left(\nabla \boldsymbol{\varphi}_j^{\mathrm{T}}\right)^{\mathrm{T}} - \nabla \cdot \boldsymbol{\varphi}_j\right\}. \tag{26}$$

Substituting (25) and (26) into (23) and then using the self-adjoint property and the identity $h_0(\boldsymbol{\phi}, \mathbf{u}, \mathbf{v}_0)\left[\boldsymbol{\varphi}_1, \boldsymbol{\varphi}_2\right] = h_0(\boldsymbol{\phi}, \mathbf{u}, \mathbf{v}_0)\left[\boldsymbol{\varphi}_2, \boldsymbol{\varphi}_1\right]$, the shape Hessian of f_0 is obtained using Proposition A.1 and A.2 in [2] as

$$\begin{aligned}
h_0(\boldsymbol{\phi}, \mathbf{u}, \mathbf{v}_0)\left[\boldsymbol{\varphi}_1, \boldsymbol{\varphi}_2\right] &= \int_{\Omega(\boldsymbol{\phi})}\left[\mathbf{S}(\mathbf{u}) \cdot \mathbf{E}(\mathbf{u})\left\{\left(\nabla \boldsymbol{\varphi}_2^{\mathrm{T}}\right)^{\mathrm{T}} \cdot \nabla \boldsymbol{\varphi}_1^{\mathrm{T}} + (\nabla \cdot \boldsymbol{\varphi}_2)(\nabla \cdot \boldsymbol{\varphi}_1)\right\}\right. \\
&\quad + \left(\nabla \mathbf{u}^{\mathrm{T}} \mathbf{S}(\mathbf{u})\right) \cdot \left\{\nabla \boldsymbol{\varphi}_1^{\mathrm{T}}\left(\nabla \boldsymbol{\varphi}_2^{\mathrm{T}}\right)^{\mathrm{T}} + \nabla \boldsymbol{\varphi}_2^{\mathrm{T}}\left(\nabla \boldsymbol{\varphi}_1^{\mathrm{T}}\right)^{\mathrm{T}}\right\} \\
&\quad \left. - 2\left(\mathbf{S}(\mathbf{u}) \mathbf{E}(\mathbf{u})\right) \cdot \left\{\nabla \boldsymbol{\varphi}_2^{\mathrm{T}}(\nabla \cdot \boldsymbol{\varphi}_1) + \nabla \boldsymbol{\varphi}_1^{\mathrm{T}}(\nabla \cdot \boldsymbol{\varphi}_2)\right\}\right] \mathrm{d}x \quad \forall \boldsymbol{\varphi}_1, \boldsymbol{\varphi}_2 \in X.
\end{aligned}$$

On the other hand, the shape Hessian of f_1 is given by

$$h_1(\boldsymbol{\phi})\left[\boldsymbol{\varphi}_1, \boldsymbol{\varphi}_2\right] = \int_{\Omega(\boldsymbol{\phi})}\left\{-\left(\nabla \boldsymbol{\varphi}_2^{\mathrm{T}}\right)^{\mathrm{T}} \cdot \nabla \boldsymbol{\varphi}_1^{\mathrm{T}} + (\nabla \cdot \boldsymbol{\varphi}_2)(\nabla \cdot \boldsymbol{\varphi}_1)\right\} \mathrm{d}x \quad \forall \boldsymbol{\varphi}_1, \boldsymbol{\varphi}_2 \in X.$$

4 Solutions to Shape Optimization Problems

Finally, we show solutions to shape optimization problems. In this section, we consider a shape optimization problem similar to Problem 3.2, but the number of constraints is changed to m. To distinguish it from Problem 3.2, we denote the problem as Problem 3.2 (m) and write its Lagrange function as

$$\mathscr{L}(\boldsymbol{\phi}, \mathbf{u}, \mathbf{v}_i) = f_0(\boldsymbol{\phi}, \mathbf{u}) + \sum_{i \in \{1,\dots,m\}} \lambda_i f_i(\boldsymbol{\phi}, \mathbf{u}). \tag{27}$$

The H^1 gradient method of domain variation type was formulated by seeking $\boldsymbol{\varphi}_{gi} \in X$ that decrease $f_i(\boldsymbol{\phi}_k, \mathbf{u})$ with respect to iterations $k \in \{0, 1, \dots\}$ by the following methods.

Problem 4.1 *(H^1 gradient method for f_i)* Let $a_X : X \times X \to \mathbb{R}$ be a bounded and coercive bilinear form in X, and c_a be a positive constant to control the magnitude of $\boldsymbol{\varphi}_{gi}$. For $\mathbf{g}_i(\boldsymbol{\phi}_k) \in X'$, find $\boldsymbol{\varphi}_{gi} \in X$ such that

$$c_a a_X\left(\boldsymbol{\varphi}_{gi}, \boldsymbol{\psi}\right) = -\left\langle \mathbf{g}_i, \boldsymbol{\psi}\right\rangle \quad \forall \boldsymbol{\psi} \in X. \tag{28}$$

As an example, we use

$$a_X(\boldsymbol{\varphi}, \boldsymbol{\psi}) = \int_{\Omega(\boldsymbol{\phi})} \left\{\left(\boldsymbol{\nabla}\boldsymbol{\varphi}^{\mathrm{T}}\right) \cdot \left(\boldsymbol{\nabla}\boldsymbol{\psi}^{\mathrm{T}}\right) + c_\Omega \boldsymbol{\varphi} \cdot \boldsymbol{\psi}\right\} \mathrm{d}x,$$

where c_Ω is a positive constant. An algorithm to solve the generalized shape optimization problem was shown in the previous paper [2] (Sect. 9.9.1 in [1]).

In addition, given the $\mathbf{g}_i$ and the computed shape Hessian h_i of f_i for all $i \in \{0, 1, \ldots, m\}$, a Newton method in H^1 class function space (H^1 Newton method) can be defined as follows (Problem 9.8.8 in [1]).

Problem 4.2 *(H^1 Newton method for domain variation type for f_i)* Let $\mathbf{g}_i \in X'$ and $h_i \in \mathfrak{L}^2(X \times X; \mathbb{R})$ for each $i \in \{0, 1, \ldots, m\}$ at $\boldsymbol{\phi}_k \in \mathscr{D}$ be given, and

$$h_{\mathscr{L}}(\boldsymbol{\phi}_k)\left[\boldsymbol{\varphi}_1, \boldsymbol{\varphi}_2\right] = h_0(\boldsymbol{\phi}_k)\left[\boldsymbol{\varphi}_1, \boldsymbol{\varphi}_2\right] + \sum_{i \in \{1,\ldots,m\}} \lambda_{ik} h_i(\boldsymbol{\phi}_k)\left[\boldsymbol{\varphi}_1, \boldsymbol{\varphi}_2\right] \quad \forall \boldsymbol{\varphi}_1, \boldsymbol{\varphi}_2 \in X$$

be the shape Hessian of $\mathscr{L}$ in (27). Moreover, let $a_X : X \times X \to \mathbb{R}$ be a bilinear form to compensate coerciveness and regularity of $h_{\mathscr{L}}(\boldsymbol{\phi}_k)$ in X, and c_a be a positive constant for adjustment. Find $\boldsymbol{\varphi}_{gi} \in X$ such that

$$h_{\mathscr{L}}(\boldsymbol{\phi}_k)\left[\boldsymbol{\varphi}_{gi}, \boldsymbol{\psi}\right] + c_a a_X\left(\boldsymbol{\varphi}_{gi}, \boldsymbol{\psi}\right) = -\left\langle g_i(\boldsymbol{\phi}_k), \boldsymbol{\psi}\right\rangle \quad \forall \boldsymbol{\psi} \in X. \tag{29}$$

A simple algorithm for solving Problem 3.2 (m) by the H^1 Newton method is shown below (Sect. 9.9.2 in [1]).

Algorithm 4.1 (The H^1 Newton method for shape optimization problems)
Obtain the local minimizer of Problem 3.2 (m) in the following way.

1. Set Ω_0 and $\boldsymbol{\phi}_0 = \mathbf{i}$ (the identity mapping) as $f_1(\boldsymbol{\phi}_0, \mathbf{u}) \leq 0$, ..., $f_m(\boldsymbol{\phi}_0, \mathbf{u}) \leq 0$. Set c_a, ε_0, ε_1, ..., ε_m appropriately. Set $k = 0$.
2. Solve the state determination problem (Problem 3.1) at $\boldsymbol{\phi}_k$ and compute $f_0(\boldsymbol{\phi}_k, \mathbf{u})$, $f_1(\boldsymbol{\phi}_k, \mathbf{u})$, ..., $f_m(\boldsymbol{\phi}_k, \mathbf{u})$. Set $I_{\mathrm{A}}(\boldsymbol{\phi}_k) = \{i \in \{1, \ldots, m\} \mid f_i(\boldsymbol{\phi}_k, \mathbf{u}) \geq -\varepsilon_i\}$.
3. Do the following when the Hessian of constraint function can be calculated.
   - Solve the adjoint problem with respect to f_0, f_{i_1}, ..., $f_{i_{|I_{\mathrm{A}}|}}$ and calculate $\mathbf{g}_0$, $\mathbf{g}_{i_1}$, ..., $\mathbf{g}_{i_{|I_{\mathrm{A}}|}}$ at $\boldsymbol{\phi}_k$.
   - Use (28) to calculate $\boldsymbol{\varphi}_{g0}$, $\boldsymbol{\varphi}_{gi_1}$, ..., $\boldsymbol{\varphi}_{gi_{|I_{\mathrm{A}}|}}$.
   - Solve $\boldsymbol{\lambda}_{k+1}$ using

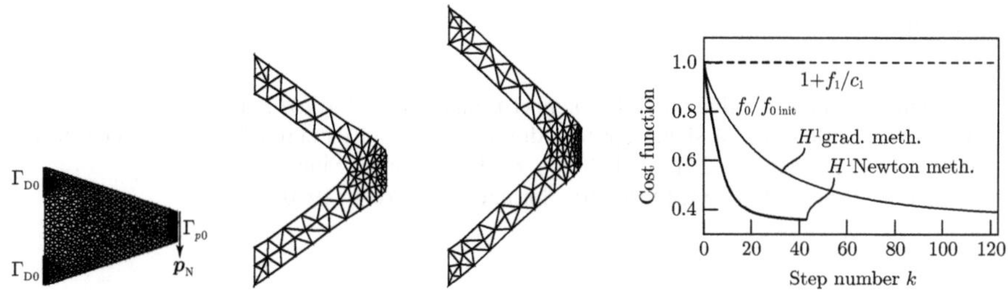

(a) Initial shape (b) H^1 grad. m. (c) H^1 Newton m. (d) Iteration history of cost functions

Fig. 1 Numerical results for two-dimensional linear elastic body

$$\left(\langle\langle \mathbf{g}_i, \boldsymbol{\varphi}_{gj}\rangle\rangle\right)_{(i,j)\in I_{\mathrm{A}}^2(\boldsymbol{\phi}_k)} \left(\lambda_{j\,k+1}\right)_{j\in I_{\mathrm{A}}(\boldsymbol{\phi}_k)} = -\left(f_i\left(\boldsymbol{\phi}_k\right) + \langle\langle \mathbf{g}_i, \boldsymbol{\varphi}_{g0}\rangle\rangle\right)_{i\in I_{\mathrm{A}}(\boldsymbol{\phi}_k)}. \quad (30)$$

If $I_{\mathrm{I}}(\boldsymbol{\phi}_k) = \{i \in I_{\mathrm{A}}(\boldsymbol{\phi}_k) \mid \lambda_{i\,k+1} < -\varepsilon_i\}$ is not empty set, replace $I_{\mathrm{A}}(\boldsymbol{\phi}_k) \setminus I_{\mathrm{I}}(\boldsymbol{\phi}_k)$ with $I_{\mathrm{A}}(\boldsymbol{\phi}_k)$ and solve (30) until $I_{\mathrm{I}}(\boldsymbol{\phi}_k) = \emptyset$.

4. Solve adjoint problems at $\boldsymbol{\phi}_k$ and compute $\mathbf{g}_0, \mathbf{g}_{i_1}, \ldots, \mathbf{g}_{i_{|I_{\mathrm{A}}|}}$ and $h_0, h_{i_1}, \ldots, h_{i_{|I_{\mathrm{A}}|}}$.
5. Solve $\boldsymbol{\varphi}_{g0}, \boldsymbol{\varphi}_{gi_1}, \ldots, \boldsymbol{\varphi}_{gi_{|I_{\mathrm{A}}|}}$ using (29).
6. Solve $\boldsymbol{\lambda}_{k+1}$ using (30). When $I_{\mathrm{I}}(\boldsymbol{\phi}_k) \neq \emptyset$, replace $I_{\mathrm{A}}(\boldsymbol{\phi}_k) \setminus I_{\mathrm{I}}(\boldsymbol{\phi}_k)$ with $I_{\mathrm{A}}(\boldsymbol{\phi}_k)$, and resolve (30) until $I_{\mathrm{I}}(\boldsymbol{\phi}_k) = \emptyset$.
7. Compute $\boldsymbol{\varphi}_g$ using

$$\boldsymbol{\varphi}_g = \boldsymbol{\varphi}_g\left(\boldsymbol{\lambda}_{k+1}\right) = \boldsymbol{\varphi}_{g0} + \sum_{i\in I_{\mathrm{A}}(\boldsymbol{\phi}_k)} \lambda_{i\,k+1}\, \boldsymbol{\varphi}_{gi}, \quad (31)$$

set $\boldsymbol{\phi}_{k+1} = \boldsymbol{\phi}_k + \boldsymbol{\varphi}_g$, and compute $f_0(\boldsymbol{\phi}_{k+1}, \mathbf{u})$, $f_1(\boldsymbol{\phi}_{k+1}, \mathbf{u})$, ..., $f_m(\boldsymbol{\phi}_{k+1}, \mathbf{u})$. Set $I_{\mathrm{A}}(\boldsymbol{\phi}_{k+1}) = \{i \in \{1, \ldots, m\} \mid f_i(\boldsymbol{\phi}_{k+1}, \mathbf{u}) \geq -\varepsilon_i\}$.
8. Assess $|f_0(\boldsymbol{\phi}_{k+1}, \mathbf{u}) - f_0(\boldsymbol{\phi}_k, \mathbf{u})| \leq \varepsilon_0$.
   - If "Yes," proceed to 9.
   - If "No," replace $k+1$ with k and return to 4.
9. Stop the algorithm.

5 Numerical Example

Figure 1 shows numerical results to Problem 3.2 obtained by the program[1] written by Kento Furuki and the author with FreeFEM++.[2]

Acknowledgements The present study was supported by JSPS KAKENHI (16K05285).

---

[1]http://www.morikita.co.jp/books/mid/061461.

[2]http://www.freefem.org/ff++/.

References

1. Azegami, H.: Shape Optimization Problems (in Japanese). Morikita Publishing, Tokyo (2016)
2. Azegami, H.: Solution of shape optimization problem and its application to product design. Mathematics for Industry 2017, vol. 26, pp. 83–98. Springer, Singapore (2016)
3. Zeidler, E.: Linear Monotone Operators. Springer, New York (1990)

Shape Optimization by Generalized J-Integral in Poisson's Equation with a Mixed Boundary Condition

Kohji Ohtsuka

Abstract Generalized J-integral is the tool for shape sensitivity analysis of singular points in boundary value problem for partial differential equations. We can solve shape optimization problems of singular points by using Generalized J-integral and H^1-gradient method (Azegami's method). Here, the mathematical method is proposed to examine shape optimization in detail by dividing the sensitivity on sets of singular points, and apply the method to Poisson's equation defined on a polygonal domain with mixed boundary condition. The boundary divides into the parts that Dirichlet boundary condition, Neumann boundary condition, and the joint of them are given. It is examined about each role of the parts of boundary in shape optimization process on a numerical example of finite element analysis.

1 Introduction

Using Generalized J-integral [8] and H^1-gradient method [2], we get the sequence of shape optimization depicted in Fig. 1 with respect to energy minimization [10]. The shape optimizations are obtained by finite element method (FEM) as follows

Step0 The initial shape Ω^0 is the unit circle (see Fig. 1a).
Step1 Get the numerical solution u_h^i with the Dirichlet condition on Γ_D^i and Neumann condition on Γ_N^i (i denotes iteration number).
Step2 Find the numerical solution $\mu^{G,i} \in W^{1,2}(\Omega^i; R^2)$ such that

$$b(\mu^{G,i}, \eta) = R_{\Omega^i}(u^i, \eta) + \int_{\Gamma_N^i} u^i (n \cdot \eta) dS \qquad \text{for all } \eta \in W^{1,2}(\Omega; R^2) \tag{1}$$

$$b(\zeta, \eta) = \int_{\Omega^i} \{\nabla\zeta : \nabla\eta + \zeta \cdot \eta\} dx, \qquad \text{(refer to (1.20) about } R_{\Omega^i}(u^i, \eta)).$$

K. Ohtsuka (✉)
Hiroshima Kokusai Gakuin University, Hiroshima, Japan
e-mail: ohtsuka@comfos.org; ohtsuka@otlab.jp

P. van Meurs et al. (eds.), *Mathematical Analysis of Continuum Mechanics and Industrial Applications II*, Mathematics for Industry 30,
https://doi.org/10.1007/978-981-10-6283-4_7

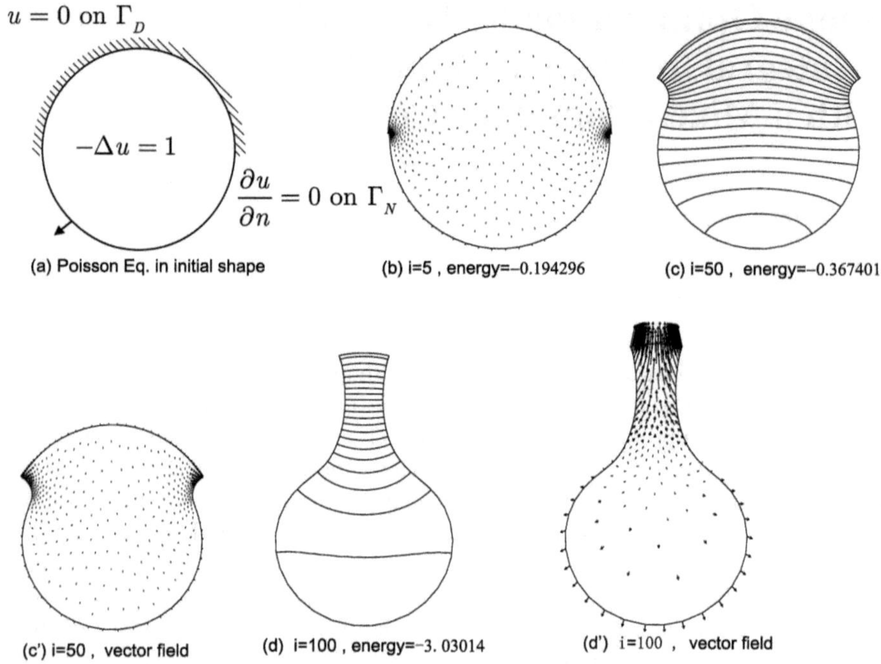

Fig. 1 Shape optimization of boundary value problem with mixed boundary condition with respect to energy

Step3 Find $\mu^{R,i}$ for the restriction condition using Lagrange multiplier. In this case, the restriction condition is the area $|\Omega^i|$ is constant.
Step4 Putting $\mu^{O,i} = \mu^{G,i} + \lambda^i \mu^{R,i}$ using the Lagrange multiplier λ^i, we have optimized shape $\Omega^{i+1} = \{x + \varepsilon\mu^{O,i}(x) : x \in \Omega^i\}$ where $\varepsilon > 0$ is small number.
Step5 If $\mu^{G,i} \simeq 0$ then Stop else goto Step1 with $\Gamma_D^{i+1} = \{x + \varepsilon\mu^{O,i}(x) : x \in \Gamma_D^i\}$ and $\Gamma_N^{i+1} = \{x + \varepsilon\mu^{O,i}(x) : x \in \Gamma_N^i\}$.

where $\nabla\zeta : \nabla\eta = \partial_k\zeta_l\partial_k\eta_l$ by Einstein summation convention, $\partial_k = \partial/\partial x_k$ and $\widehat{W}(u) = |\nabla u|^2/2$. To find optimized shape by (1) is called H^1-gradient method [2]. The integral $R_{\Omega^i}(u^i, \eta)$ is the part of Generalized J-integral. The numerical calculation is done by FreeFem++ [5].

In this paper, we study mathematically to elucidate why such shapes are provided. Generalized J-integral (18) is a mathematical tool to measure sensitivity of singular points of solutions. However, FEM is calculated in the polygon (polyhedral) domain, so that various singularities will appear. Then, we prepare for mathematical results on domains with local Lipschitz condition.

2 Sobolev Spaces on Domains with Local Lipschitz Condition

A bounded open set D in $R^d(d = 2, 3)$ is called a domain with local Lipschitz property if there exist positive numbers α, β, and M, a locally finite open cover $\{U_r\}_{r=1,\cdots,N}$ of ∂D, and for each r, Lipschitz continuous function a_r of $d-1$ variables, and they satisfy the following conditions for some Cartesian coordinate system $(\xi_{r,1}, \cdots, \xi_{r,d})$ in U_r

1. Each function a_r satisfies a Lipschitz condition with constant M: that is,

$$
\begin{aligned}
&U_r' = \{\xi_r' = (\xi_{r,1}, \cdots, \xi_{r,d-1}) \in R^{d-1} : |\xi_{r,i}| < \alpha, i = 1, \cdots, d-1\} \\
&|a_r(\xi') - a_r(\eta')| \le M|\xi' - \eta'| \quad \text{for all } \xi', \eta' \in U_r' \\
&U_r = \{\xi_r \in R^d : a_r(\xi_r') - \beta < \xi_{r,d} < a_r(\xi_r') + \beta \quad \text{for all } \xi_r' \in U_r'\}
\end{aligned}
$$

2. $\partial D \cap U_r$ and $D \cap U_r$ are represented by

$$\xi_{r,d} = a_r(\xi_r') \text{ and } a_r(\xi_r') < \xi_{r,d} < a_r(\xi_r') + \beta, \quad \text{for all } \xi_r' \in U_r'$$

For an integer $m \ge 0$ and $0 \le p \le \infty$, let $W^{m,p}(D)$ be Sobolev space of p-integrable functions which have p-integrable derivatives of m-th order in D with the norm $\|\cdot\|_{m,p,D}$ (see e.g., [1, 3.1]), where

$$\|v\|_{m,\infty,D} = \max_{0\le|\alpha|\le m} \operatorname{ess\,sup}_{x\in D} |D^\alpha v(x)|$$

For an integer m and $s = m + \sigma$, $0 < \sigma < 1$, Sobolev space $W^{s,p}(D)$ is defined as the space of functions with the norm (see e,g., [1, 7.48])

$$\|v\|_{s,p,D} = \left\{ \|v\|^p_{m,p,D} + \sum_{|\alpha|=m} \int_D \int_D \frac{|\nabla^\alpha v(x) - \nabla^\alpha v(y)|^p}{|x-y|^{d+\sigma p}} dxdy \right\}^{1/p}$$

The surface integral of the function f over ∂D is written using the partition of unity η_r, $r = 1, \cdots, N$ for ∂D subordinate to $\{U_r\}$ (see e.g., [1, 3.14]) as

$$\int_{\partial D} f\,dS = \sum_{r=1}^N \int_{U_r'} \eta_r f(\xi_r', a_r(\xi_r'))dS_r(\xi_r') \tag{2}$$

$$dS_r(\xi_r') = \left\{1 + |\nabla' a_r(\xi_r')|^2\right\}^{1/2} d\xi_r', \quad |\nabla' a_r(\xi_r')|^2 = \sum_{i=1}^{d-1} (\partial a_r(\xi_r')/\partial \xi_{r,i})^2$$

Sobolev spaces $W^{s,p}(\partial D)$ are defined similarly.

The following results hold for a domain D with local Lipschitz property.

1. (Density theorem) $C^\infty(\overline{D})$ is dense in $W^{m,p}(D)$ for the integer $m \geq 0$ and $1 \leq p < \infty$ (see e.g., [1, 3.18]).
2. (Trace theorem 1) For any $v \in C^\infty(\overline{D})$ and $1 \leq p < \infty$, the trace $\gamma_0 : v \mapsto v(x'), x' \in \partial D$ satisfies the following estimation

$$\|\gamma_0 v\|_{0,q,\partial D} \leq C\|v\|_{1,p,D} \quad \text{with a constant } C > 0 \text{ independent of } v \tag{3}$$

   where $q = 1/p - [1/(d-1)](p-1)/p$ if $1 \leq p < d$ (see [7, Theorem 2.4.2]), arbitrary $q \geq 1$ if $p \geq d$ (see [7, Theorem 2.4.6] and [7, 2.3.8])
3. (Trace theorem 2) The trace $\gamma_0 : v \mapsto v(x'), x' \in \partial D$ is the bounded continuous operator from $W^{s,p}(D)$ onto $W^{s-1/p,p}(\partial D)$ for $1 < p < \infty$ if $1/p < s \leq 1$ (see e.g., [1, 7.53]). Notice that Trace theorem 2 does not hold when $p = 1, s = 1$, but (3) holds for $p = 1, q = 1$.

Theorem 2.1 (Approximation of domain) *There is the number $h_0 > 0$ and domains $D^h, 0 < h < h_0$ with smooth boundary such that $\overline{D^h} \subset D$ whose boundary $\partial D^h \cap U_r$ and $D^h \cap U_r$ are expressed as*

$$\xi_{r,d} = a_r^h(\xi_r') \text{ and } a_r^h(\xi_r') < \xi_{r,d} < a_r^h(\xi_r') + \beta, \quad \text{for all } \xi_r' < U_r'$$

They satisfy conditions for each $r = 1, \cdots, N$;

$$\overline{D^{h_2}} \subset D^{h_1} \quad \text{if } h_1 < h_2, \text{ and } \lim_{h \to 0} D^h = D \tag{4}$$

$$|a_r^h(\xi') - a_r^h(\eta')| \leq M|\xi' - \eta'| \quad \text{for all } 0 < h < h_0, \quad \text{and } \xi', \eta' \in U_r' \tag{5}$$

$$\lim_{h \to 0}\left(\sup_{\xi_r' \in U_r'} |a_r^h(\xi_r') - a_r(\xi_r')|\right) = 0 \tag{6}$$

$$\lim_{h \to 0} \int_{U_r'} \left|\frac{\partial a_r}{\partial \xi_{r,i}} - \frac{\partial a_r^h}{\partial \xi_{r,i}}\right|^p d\xi' = 0 \quad \text{for } 1 \leq p < \infty, i = 1, \cdots, d-1 \tag{7}$$

For the proof we refer to [6].

Lemma 2.2 *Let $\Omega^h, h > 0$ be the approximation of domain Ω given in Theorem 2.1. For any $v \in W^{s,p}(D), s > 1/p$ and for any $r = 1, \cdots N$,*

$$\lim_{h \to 0} \int_{U_r'} |v(\xi_r', a_r^h(\xi_r')) - v(\xi_r', a_r(\xi_r'))|^p d\xi_r' = 0 \tag{8}$$

Proof. Let $v \in C^\infty(\overline{D})$. From (5), for any $\varepsilon > 0$, there is a number $h(\varepsilon) > 0$ such that

$$a_r(\xi_r') < a_r^h(\xi_r') < a_r(\xi_r') + \varepsilon \quad \text{for all } \xi_r' \in U_r, h \le h(\varepsilon)$$

$$|v(\xi', a_r^h(\xi_r')) - v(\xi', a_r(\xi_r'))| \le \int_{a_r(\xi_r')}^{a_r^h(\xi_r')} \left| \frac{\partial v}{\partial \xi_{r,d}}(\xi_r', \eta) \right| d\eta$$
$$\le \int_{a_r(\xi_r')}^{a_r(\xi_r')+\varepsilon} \left| \frac{\partial v}{\partial \xi_{r,d}}(\xi_r', \eta) \right| d\eta$$

for all $h < h(\varepsilon)$ which leads by Hölder's inequality that

$$|v(\xi', a_r^h(\xi_r')) - v(\xi', a_r(\xi_r'))|^p \le \varepsilon^{p-1} \int_{a_r(\xi_r')}^{a_r(\xi_r')+\varepsilon} \left| \frac{\partial v}{\partial \xi_{r,d}}(\xi_r', \eta) \right|^p d\eta$$

Then, we have the estimation

$$\int_{U_r'} |v(\xi_r', a_r^h(\xi_r')) - v(\xi_r', a_r(\xi_r'))|^p d\xi' \le C_1 \varepsilon^{p-1} \|v\|_{1,p,U_r}^p$$

with a constant $C_1 > 0$ independent of v. Therefore, the trace operators $\gamma_0^h v(\xi') = v(\xi', a_r^h(\xi'))$, $\gamma_0 v(\xi') = v(\xi', a_r(\xi'))$, $\xi' \in U_r'$ satisfy

$$\|(\gamma_0^h - \gamma_0)v\|_{0,p,U_r'} \le C_2(\varepsilon^{p-1})^{1/p} \|v\|_{1,p,U_r} \tag{9}$$

The space $W^{s,p}(D)$ is the interpolation between $W^{1,p}(D)$ and $L^p(D)$, that is, $W^{s,p}(D) = T(p, 1-s-\frac{1}{p}, W^{1,p}(D), L^p(D))$ (see [1, 7.36]). Take the number $s_0 > 1/p$ such that $s_0 < s$. Then $W^{s,p}(D) = T(p, 1-\frac{s-s_0}{1-s_0} - \frac{1}{p}, W^{1,p}(D), W^{s_0,p}(D))$. By Trace theorem 2, we have

$$\|(\gamma_0^h - \gamma_0)v\|_{0,p,U_r'} \le C_3 \|v\|_{s_0,p,U_r} \tag{10}$$

Using the interpolation theorem [1, 7.17], we arrive from (9) and (10)

$$\|(\gamma_0^h - \gamma_0)v\|_{0,p,U_r'} \le \left(C_2 \varepsilon^{(p-1)/p}\right)^{1-\theta} C_3^\theta \|v\|_{s,p,U_r}$$

where $\theta = (1-s)/(1-s_0)$. Since $1-\theta = 1-\frac{1-s}{1-s_0} = (s-s_0)/(1-s_0) > 0$, we can prove (8) by Density theorem. □

Lemma 2.3 *Let $\Omega^h, h > 0$ be the approximation of domain Ω given in Theorem2.1. For any $f \in W^{s,p}(\Omega), g \in W^{s,q}(\Omega)$, $p > 1$, $p^{-1} + q^{-1} = 1$, $1/p < s \le 1$, we have*

$$\lim_{h \to 0} \int_{\partial D^h} (\gamma_0^h f)(\gamma_0^h g) dS^h = \int_{\partial D} (\gamma_0 f)(\gamma_0 g) dS \tag{11}$$

Proof. The surface integral on ∂D^h is written from (2) as

$$\int_{\partial D} (\gamma_0^h f)(\gamma_0^h g)\, dS^h = \int_{U_r'} \eta_r (\gamma_0^h f)(\gamma_0^h g)(1 + |\nabla' a_r^h|^2)^{1/2} d\xi_r'$$

During the proof, we will omit the index r. By Trace theorem 2, for the surface element, we have

$$1 + |\nabla' a_r^h|^2 = \left(1 + |\nabla' a_r|^2\right)\left(1 + \frac{|\nabla' a_r^h|^2 - |\nabla' a_r|^2}{1 + |\nabla' a_r|^2}\right)$$

$$\sqrt{1 + |\nabla' a_r^h|^2} - \sqrt{1 + |\nabla' a_r|^2} = \sqrt{1 + |\nabla' a_r|^2}\left\{\sqrt{1 + \frac{|\nabla' a_r^h|^2 - |\nabla' a_r|^2}{1 + |\nabla' a_r|^2}} - 1\right\}$$

$$\leq \frac{1}{2}\left||\nabla' a_r^h|^2 - |\nabla' a_r|^2\right| \left(1 + |\nabla' a_r|^2\right)^{-1/2}$$

$$\leq M \sum_{i=1}^{d-1} \left|\partial a^h/\partial \xi_i - \partial a/\partial \xi_i\right| \tag{12}$$

where we used that $\operatorname{ess\,sup}_{\xi' \in U'} |\partial a^h(\xi')/\partial \xi_i| \leq M$ and $\sqrt{1+c} \leq 1 + \frac{1}{2}c$ if $c > 0$. By Trace theorem 2, there is a number $p' > p$ such that $\gamma_0 f \in L^{p'}(U')$. For $i = 1, \cdots, d-1$, by Hölder's inequality we obtain

$$\int_{U'} (\gamma_0 f)(\gamma_0 g) \left|\frac{\partial a^h}{\partial \xi_i} - \frac{\partial a}{\partial \xi_i}\right| dS \leq \left\|\gamma_0 f \left|\frac{\partial a^h}{\partial \xi_i} - \frac{\partial a}{\partial \xi_i}\right|\right\|_{0,p,U'} \|\gamma_0 g\|_{0,q,U'}$$

$$\left\|\gamma_0 f \left|\frac{\partial a^h}{\partial \xi_i} - \frac{\partial a}{\partial \xi_i}\right|\right\|_{0,p,U'}^p \leq \|\gamma_0 f\|_{0,p',U'}^p \left\{\int_{U'} \left|\frac{\partial a^h}{\partial \xi_i} - \frac{\partial a}{\partial \xi_i}\right|^{\frac{pp'}{p'-p}} d\xi'\right\}^{\frac{p'-p}{p}}$$

From the estimation just above, (8), (7), and (12), we have

$$\int_{U'} (\gamma_0^h f)(\gamma_0^h g) dS^h - \int_{U'} (\gamma_0 f)(\gamma_0 g) dS = \int_{U'} (\gamma_0^h f - \gamma_0 f)(\gamma_0^h g) dS^h$$
$$+ \int_{U'} (\gamma_0 f)(\gamma_0^h g - \gamma_0 g) dS^h + \int_{U'} (\gamma_0 f)(\gamma_0 g)(dS^h - dS) \to 0 \quad \text{as } h \to 0$$

This complete the proof of (11). □

Lemma 2.4 *Let $\Omega^h, h > 0$ be the approximation of domain Ω given in Theorem 2.1. For any $f \in W^{s,p}(\Omega), g \in W^{s,q}(\Omega)$, $p > 1, p^{-1} + q^{-1} = 1, 1/p < s \leq 1$, we have for $k = 1, \cdots, d$*

$$\lim_{h\to 0}\int_{\partial D^h}(\gamma_0^h f)(\gamma_0^h g)n_k^h dS^h = \int_{\partial D}(\gamma_0 f)(\gamma_0 g)n_k dS \tag{13}$$

where n, n^h are the outward unit normals of ∂D, ∂D^h, respectively.

Proof. Since $n^h(x) = (1+|\nabla' a_r^h|^2)^{-1/2}(\partial a_r^h/\partial\xi_1, \cdots, -1)$ if $x \in U_r$, we have for $k = 1, \cdots, d-1$,

$$\int_{\partial D^h}(\gamma_0^h f)(\gamma_0^h g)n_i^h dS^h = \sum_{r=1}^{N}\int_{\partial U_r'}\eta_r(\gamma_0^h f)(\gamma_0^h g)(\partial a_r^h/\partial\xi_k)d\xi'$$

We can prove (13) if $k < d$ similarly as in Lemma 2.3. It is easy to prove (13) when $k = d$. □

3 Shape Sensitivity by Generalized J-Integral

Let Ω be a polygonal domain whose boundary is not smooth but satisfies local Lipschitz condition. Let us denote the each segment of $\partial\Omega$ by $\overline{\Gamma_j}$, $j = 1\cdots, N$ where Γ_j is open. The segments are numbered in such a way that Γ_{j+1} follows Γ_j according to the positive orientation. We denote by γ_j the vertex which is the endpoint of Γ_j and by Θ_j the measure of the interior angle at γ_j.

We now consider the boundary value problem: For a given $f \in W^{1,2}(R^2)$, find a $u(\Omega)$ such that

$$-\Delta u(\Omega) = f \quad \text{in } \Omega, \tag{14}$$

$$u(\Omega) = 0 \quad \text{on } \Gamma_D, \qquad \partial u(\Omega)/\partial n = 0 \quad \text{on } \Gamma_N \tag{15}$$

where $\Gamma_D = \sum_{j=1}^{N_D}\Gamma_j$, $\Gamma_N = \sum_{j=N_D+1}^{N}\Gamma_j$ with a integer $0 < N_D \le N$. If $1 < N_D < N$, then there is unique solution $u(\Omega)$ and has singularity at γ_j such as

$$\text{if } j \ne \gamma_{N_D}, \gamma_N, \text{ and } \Theta_j > \pi, \quad u(\Omega) = K_j r_j^{\pi/\Theta_j} S_j(x) + u_{j,R} \quad \text{near } \gamma_j$$

$$\text{if } j = \gamma_{N_D}, \gamma_N, \text{ and } \Theta_j > \pi/2, \quad u(\Omega) = K_j r_j^{\pi/2\Theta_j} S_j(x) + u_{j,R} \quad \text{near } \gamma_j$$

where for each j, K_j stands for a constant, $r_j = |x - \gamma_j|$, S_j is smooth function and $u_{j,R} \in W^{2,2}(\text{near } \gamma_j)$. At $j = \gamma_{N_D}, \gamma_N$, if $\Theta_j > \pi/2$, then $u(\Omega) \in W^{1+\alpha,2}(\Omega)$, $0 < \alpha < 1/2$. But $u(\Omega) \in W^{1+\beta,2}(\Omega \setminus \overline{B_{DN,\varepsilon}})$, $\beta > 1/2$ with $B_{DN,\varepsilon} = B_\varepsilon(\gamma_{N_D}) \cup B_\varepsilon(\gamma_N)$, $B_\varepsilon(\gamma_j) = \{x : |x-\gamma_j| < \varepsilon\}$ for small number $\varepsilon > 0$. If $N_D = N$, then $u \in W^{1+\alpha,2}(\Omega)$, $\alpha > 1/2$ (see e.g., [4]).

Let $u_h(\Omega)$ be the finite element approximation using P^1-element under the mesh $\mathcal{T}(\Omega P)$ of Ω with size $h > 0$. The projection Π_h from $C^\infty(\overline{\Omega})$ onto the finite element space $V_h(\mathcal{T}(\Omega_1, P^1)$ has the estimation

$$\|v - v_h\|_{1,2,\Omega} \leq C_1 h \|v\|_{2,2,\Omega}; \quad \|v - v_h\|_{1,2,\Omega} \leq C_2 \|v\|_{1,2,\Omega}$$

Then, we apply the interpolation theorem [1, 7.17] to $(1 - \Pi_h)$, and we arrive the estimation for $v \in W^{1+\alpha,2}(\Omega)$

$$\|v - v_h\|_{1,2,\Omega} \leq C_3 h^\alpha \|v\|_{1+\alpha,2,\Omega} \tag{16}$$

with a constant $C_3 > 0$ independent of h. By Céa's lemma (see e.g., [3, Lemma 2.28], the following hold

$$\|u_h(\Omega) - u(\Omega)\|_{1,2,\Omega} \leq C_4 h^\alpha \|u(\Omega)\|_{1+\alpha,2,\Omega} \tag{17}$$

with a constant $C_3 > 0$ independent of h.

3.1 Generalized J-Integral

For $u = u(\Omega)$, we now introduce *GJ-integral* $J_\omega(u(\Omega), \mu)$ with two parameters; a domain ω and a vector field $\mu \in W^{1,\infty}(R^d; R^d)$

$$J_\omega(u, \mu) = P_\omega(u, \mu) + R_\omega(u, \mu), \quad \widehat{W}(u) = \frac{1}{2}|\nabla u|^2 \tag{18}$$

$$P_\omega(u, \mu) = \int_{\partial(\omega\cap\Omega)} \left\{ \widehat{W}(u)(\mu \cdot n) - \frac{\partial u}{\partial n}(\nabla u \cdot \mu) \right\} dS \tag{19}$$

$$R_\omega(u, \mu) = -\int_{\omega\cap\Omega} \left\{ f(\nabla u \cdot \mu) - (\nabla u \cdot \nabla \mu_k)\frac{\partial u}{\partial x_k} + \widehat{W}(u)\mathrm{div}\mu \right\} dx \tag{20}$$

By (17) and Schwarz inequality, we have the estimation

$$\int_\Omega (|\nabla u|^2 - |\nabla u_h|^2) dx \leq \left\{ \int_\Omega (|\nabla u| + |\nabla u_h|)^2 dx \right\}^{1/2} \|u - u_h\|_{1,2,\Omega} \tag{21}$$

from which $|R_\omega(u_h, \mu) - R_\omega(u, \mu)| \leq C h^\alpha \|u\|_{1+\alpha,2,\Omega}$ with a constant $C > 0$ independent of h.

Important property of GJ-integral is to take zero if $u(\Omega)|_{\omega\cap\Omega} \in W^{2,2}(\omega \cap \Omega)$ [8, Theorem 3.4]. Furthermore, we get the following.

Theorem 3.1 *If $\omega \cap \Omega$ has local Lipschitz property and $u(\Omega)|_{\omega\cap\Omega} \in W^{1+\alpha,2}(\omega \cap \Omega)$, $\alpha > 1/2$, then GJ-integral is well-defined and satisfies*

$$J_\omega(u, \mu) = 0 \quad \textit{for all } \mu \in W^{1,\infty}(R^2; R^2) \tag{22}$$

Proof. We have the smooth domains $D^h, h > 0$ which approximate $\omega \cap \Omega$ (see Theorem 2.1). Since $\overline{D^h} \subset \omega \cap \Omega$, we have $J D^h(u, \mu) = 0$ by interior regularity of

$u \in W^{2,2}(D^h)$. It is easy to show that $\lim_{h\to 0} R_{D^h}(u, \mu) = R_\omega(u, \mu)$. Also we have the convergence

$$P_{D^h}(u, \mu) = \int_{\partial D^h} \left\{\gamma_0^h \widehat{W}(u)(\mu \cdot (n^h - n)) - (n^h - n) \cdot \gamma_0^h(\nabla u)(\gamma_0^h(\nabla u) \cdot \mu)\right\} dS^h$$
$$+ \int_{\partial D^h} \left\{\gamma_0^h \widehat{W}(u)(\mu \cdot n) - n \cdot \gamma_0^h(\nabla u)(\gamma_0^h(\nabla u) \cdot \mu)\right\} dS^h$$

where γ_0^h stands for the trace operator from $W^{s,p}(D^h)$ onto $W^{1,s-1/p}(\partial D^h)$. From Lemmas 2.3 and 2.4, we can prove $P_{D^h}(u.\mu) \to P_\omega(u, \mu)$ as $h \to 0$. Therefore, $0 = J_{D^h}(u, \mu) \to J_\omega(u, \mu)$ as $h \to 0$, which leads (22). □

For Γ_0, we denote by $P_\omega(u, \mu|\Gamma_0)$ the integral

$$\int_{\Gamma_0 \cap \partial(\omega \cap \Omega)} \left\{\widehat{W}(x, u)(\mu \cdot n) - \frac{\partial u}{\partial n}(\nabla u \cdot \mu)\right\} dS \tag{23}$$

Next, we introduce *JO-integral* $J_\omega^O(u, \mu) = P_\omega(u, \mu|\partial\omega \cap \Omega) + R_\omega(u, \mu)$ that was nearer to *original* J-integral than $J_\omega(u, \mu)$ from a physical point of view. It is clear that $(\omega \cap \partial\Omega) \cap (\partial\omega \cap \Omega) = \emptyset$.

3.2 Shape Optimization with Respect to Energy

In this subsection, we analyze the situation of Fig. 1. We now consider the perturbation $x \mapsto \varphi_{t,\mu}(x) = x + t\mu(x)$, $\mu \in W^{1,\infty}(R^2; R^2)$. The shape optimization is given by $\Omega^{i+1} = \varphi_{\varepsilon,\mu^{O,i}}(\Omega^i)$ for $i = 0, 1, \cdots$. Putting $\Omega = \Omega^i$, $\Gamma_D = \Gamma_D^i$, and $\Omega(t) = \varphi_t(\Omega)$, $\Gamma_D(t) = \varphi_t(\Gamma_D)$ with $\varphi_t = \varphi_{t,\mu}$ for arbitrary $\mu \in W^{1,\infty}(R^2; R^2)$, we define the perturbation $u(t)$ of $u = u(\Omega^i)$ as follows

$$\mathscr{E}(u(t); 1, \Omega(t)) = \min_{v \in V(\Omega)^{\varphi_t}} \mathscr{E}(v; 1, \Omega(t)) \tag{24}$$
$$\mathscr{E}(v; 1, \Omega(t)) = \int_{\Omega(t)} \left(\frac{1}{2}|\nabla v|^2 - u\right) dx$$
$$V(\Omega)^{\varphi_t} = \{v \in W^{1,2}(\Omega(t)) : \quad v = 0 \quad \text{on } \Gamma_D(t)\}$$

Theorem 3.2 ([8, 9]) *For the solution $u(t)$ of Problem (24) we have for $\overline{\Omega} \subset \omega$*

$$\left.\frac{d}{dt}\mathscr{E}(u(t); 1, \Omega(t))\right|_{t=0} = -J_\omega^O(u, \mu) - \int_{\Gamma_N} u(\mu \cdot n)\, dS \tag{25}$$

Here we notice that $J_\omega^O(u, \mu) = R_\Omega(u, \mu)$.

Let us denote $\gamma_{D,k}^{i+1} = \varphi_{\varepsilon,\mu^{O,i}}(\gamma_{D,k}^i)$, $k = 1, 2$ with $\gamma_{D,1}^0 = (1, 0)$, $\gamma_{D,2}^0 = (-1, 0)$, $u^i = u(\Omega^i)$, $B_\varepsilon(\gamma_{D,k}^i) = \{x \in R^2 : |x - \gamma_{D,k}^i| < \varepsilon\}$, and $\Omega_\varepsilon^i = \Omega^i \setminus \cup_{k=1}^2 \overline{B_\varepsilon(\gamma_{D,k}^i)}$. Then $J_{\Omega_\varepsilon^i}(u^i, \mu) = 0$ by Theorem 3.1 for any $\mu \in W^{1,\infty}(R^2; R^2)$, it follows that

$$
\begin{aligned}
-R_{\Omega^i}(u^i, \mu) &= -R_{\Omega_\varepsilon^i}(u^i, \mu) - \sum_{k=1}^{2} R_{B_\varepsilon(\gamma_{D,k}^i)}(u^i, \mu) \\
&= P_{\Omega_\varepsilon^i}(u^i, \mu|\Gamma_D^i) + P_{\Omega_\varepsilon^i}(u^i, \mu|\Gamma_N^i) \qquad (26)\\
&\quad + \sum_{k=1}^{2} P_{\Omega_\varepsilon^i}(u^i, \mu|\partial B_\varepsilon(\gamma_{D,k}^i)) - \sum_{k=1}^{2} R_{B_\varepsilon(\gamma_{D,k}^i)}(u^i, \mu) \qquad (27)
\end{aligned}
$$

From Theorem 3.2 and the relation

$$
J_{B_\varepsilon(\gamma_{D,k}^i)}^O(u^i, \mu) = -P_{\Omega_\varepsilon^i}(u^i, \mu|\partial B_\varepsilon(\gamma_{D,k}^i)) + R_{B_\varepsilon(\gamma_{D,k}^i)}(u^i, \mu)
$$

we arrive that

$$
\begin{aligned}
\left.\frac{d}{dt}\mathscr{E}(u^i(t); 1, \Omega^{\varphi_t})\right|_{t=0} &= P_{\Omega_\varepsilon^i}(u^i, \mu|\Gamma_D^i) + P_{\Omega_\varepsilon^i}(u^i, \mu|\Gamma_N^i) \\
&\quad - \sum_{k=1}^{2} J_{B_\varepsilon(\gamma_{D,k}^i)}^O(u^i, \mu) - \int_{\Gamma_N} u^i(\mu \cdot n)\, dS \qquad (28)
\end{aligned}
$$

$$
P_{\Omega_\varepsilon^i}(u^i, \mu|\Gamma_D) = -\int_{\Gamma_{D,\varepsilon}^i} \frac{1}{2}\left|\frac{\partial u^i}{\partial n}\right|^2 (\mu \cdot n) dS, \quad \Gamma_{D,\varepsilon}^i = \Gamma_D^i \setminus \overline{B_\varepsilon(\gamma_D^i)}
$$

$$
P_{\Omega_\varepsilon^i}(u^i, \mu|\Gamma_N) = \int_{\Gamma_{N,\varepsilon}^i} \frac{1}{2}|\nabla u^i|^2 (\mu \cdot n) dS, \quad \Gamma_{N,\varepsilon}^i = \Gamma_N \setminus \overline{B_\varepsilon(\gamma_D^i)}
$$

where we put $B_\varepsilon(\gamma_D^i) = \cup_{k=1}^2 B_\varepsilon(\gamma_{D,k}^i)$ and used that $\nabla u = \partial u/\partial n$ on Γ_D. The angle between Γ_D^0 and Γ_N^0 is near π, so that $u^0 = K_k r_k^{1/2} S_k + u_R$ near $\gamma_{D,k}^0$, $k = 1, 2$, hence

$$
\lim_{\varepsilon \to 0} J_{B_\varepsilon(\gamma_{D,k}^0)}(u^0, \mu) = \frac{\pi}{8} K_k^2 \mu(\gamma_{D,k}^0) \cdot \tau_{D,k}(\gamma_{D,k}^0)
$$

where $\tau(\gamma_{D,k}^0)$ is the tangential vector on $\partial\Omega$ at $\gamma_{D,k}^0$ in the direction of Γ_D^0 from Γ_N^0.

We observe three stages in Fig. 1:

Stage1: $J_{B_\varepsilon(\gamma_{D,k}^i)}(u^i, \mu^{O,i}) \to 0$ as $i = 1, 2, \cdots, 5$ because the interior angle $\Theta_{D,k}^5 < \pi$, $k = 1, 2$ between Γ_D^5 and Γ_N^5 and $\Theta_{D,k}^{50} < \pi/2$ (see Fig. 1b).

Stage2: $P_\Omega(u^{i+1}, \mu^{O,i+1}|\Gamma_D^{i+1}) < P_\Omega(u^i, \mu^{O,i}|\Gamma_D^i)$ by $|\Gamma_D^{i+1}| < |\Gamma_D^i|$. $P_\Omega(u^{i+1}, \mu^{O,i+1}|\Gamma_N^{i+1}) < P_\Omega(u^i, \mu^{O,i}|\Gamma_N^i)$ by $|\nabla u^{i+1}| < |\nabla u^i|$, this means u^i become constant as $i \to \infty$ (see Fig. 1c and c').

Stage3: For $i \geq 100$, let $\Omega_c^i = \{x \in \Omega^i : u^i(x) \geq c\}$ where c is constant. There will be a monotonically increasing sequence $c_i \to \infty$ as $i \to \infty$ such that $|\Omega_{c_i}^i| > \pi/2$ by the observation that the term $-\int_{\partial\Gamma_N^i} u^i(\mu^{O,i} \cdot n)dS \to -\infty$ as $i \to \infty$. The Dirichlet part Γ_D^i become very small. The remaining domain $\Omega^i \setminus \overline{\Omega_{c_i}^i}$ tying Γ_D^i become narrow bands to lower influence of $|\nabla u^i|$ (see Fig. 1d and d').

Acknowledgements I am deeply grateful to Prof. H. Azegami and M. Kinumra. This work was supported by JSPS KAKENHI Grant Number 16K05285.

References

1. Adams, R.A.: Sobolev Spaces. Academic Press (1975)
2. Azegami, H., Wu, Z.: Domain optimization analysis in linear elastic problems: approach using traction method. JSME Int. J. Ser. A **39**(2), 272–278 (1996)
3. Ern, A., Guermond, J.-J.: Theory and Practice of Finite Elements. Springer (2004)
4. Grisvard, P.: Singularities in Boundary Value Problems. Springer (1992)
5. Hecht, F.: New development in FreeFem++. J. Numer. Math. **20**, 251–265 (2012)
6. Nečas, J.: Sur les domaines du type N. Czechoslov. Math. J. **12**, 285–287 (1962)
7. Nečas, J.: Les méthodes directes en théorie des équations elliptoques, Academia, Praha, and Masson et Cie, Editeurs, Paris (1967) (English translation: Direct Methods in the Theory of Elliptic Equations. Springer, 2012)
8. Ohtsuka, K.: Generalized J-integral and its applications. I.-Basic theory. Jpn. J. Appl. Math. **2**, 329–350 (1985)
9. Ohtsuka, K., Khludnev, A.: Generalized J-integral method for sensitivity analysis of static shape design. Control Cybern. **29**, 513–533 (2000)
10. Ohtsuka, K.: Mathematical theory on perturbation of singular points in continuum mechanics and its application to fracture and shape optimization, Math. Ind. Res. **2**: 203–252 (2015) (Kyushu Univ)

Part III
Modeling of Earthquakes

On Applications of Fast Domain Partitioning Method to Earthquake Simulations with Spatiotemporal Boundary Integral Equation Method

Ryosuke Ando

Abstract This paper introduces the recent developments in the earthquake rupture simulations particularly focusing on our applications of the spatiotemporal domain boundary integral equation method (ST-BIEM) and the fast domain partitioning method (FDPM), which enable us to reduce the required memory storage and the computation time, respectively, to $O(M^2)$ and $O(MN^2)$ from the original values of $O(M^2N)$ and $O(MN^3)$ for the given elements M and time steps N. FDPM utilizes the particular spatiotemporal dependence of the stress Green's function (fundamental solutions) by partitioning the causality cone. FDPM can also seamlessly combine fully dynamic and quasi-dynamic simulation algorithms adapted in seismology. Related issues in seismological simulations are also discussed.

1 Introduction

This article together with Chap. 9 is intended to provide a minimum overview and technical notes of the spatiotemporal boundary integral equation method (ST-BIEM) applied to dynamically propagating rapture problems such as earthquakes. BIEM has utility in seismology due to its applicability to the complicated geometry of natural fault systems [6, 28] retaining the numerical accuracy in the stress singularity analysis, which is crucial for the evaluation of fracture criterion. BIEM also suits for the seamless connections of the quasi-dynamic and fully dynamic processes such as nucleation of fractures or recurrence of fracture events, both of which have particular importance in understanding earthquake generation. We can see ST-BIEM become to be applied for large scale problems by recently increasing efficiency due to the fast domain partitioning method (FDPM) [2].

BIEM is a semi-analytic method optimal for seismology dealing with strain as small as 10^{-4}, used to solve the elastic response described by a linear partial derivative

R. Ando (✉)
Department of Earth and Planetary Science, School of Science, University of Tokyo, 7-3-1 Hongo, Bunkyo-ku, Tokyo 113-0033, Japan
e-mail: ando@eps.s.u-tokyo.ac.jp

P. van Meurs et al. (eds.), *Mathematical Analysis of Continuum Mechanics and Industrial Applications II*, Mathematics for Industry 30,
https://doi.org/10.1007/978-981-10-6283-4_8

equation (the Nervier equation) $\ddot{\boldsymbol{u}} = \boldsymbol{f}/\rho + \alpha^2\nabla(\nabla \cdot \boldsymbol{u}) - \beta^2\nabla \times (\nabla \times \boldsymbol{u})$, where $\boldsymbol{u}$, $\boldsymbol{f}$ and ρ respectively denote the displacement, the body force, and the density. α and β, respectively, denote P (dilatational) and S (shear) wave speeds given as $\alpha = \sqrt{(\lambda + 2\mu)/\rho}$ and $\beta = \sqrt{\mu/\rho}$ with the Lame constants λ and μ. In BIEM, the elastic field is described in the form of the boundary integral equations (BIEs) as $u(\boldsymbol{x}, t) = \int\int \Delta u(\boldsymbol{\xi}, \tau) K(\boldsymbol{x}, t - \tau; \boldsymbol{\xi}, 0)\, d\tau d\boldsymbol{\xi}$, which describe spatiotemporal convolutions of the Green's function K (the integration kernel or the fundamental solution) and the slip history Δu. The applications of BIEM span various aspects of the earthquake generations, including the dynamic rupture of earthquakes, stress buildup (e.g., [4, 21, 33]), gradual nucleation (e.g., [25]), and formation of faults (e.g., [3]).

In the following, I briefly outline the recent history of the BIEM development and relevant backgrounds in seismology for the readers in the other field. The elasto-dynamic BIEM was first seismologically applied to the case of the 2-D anti-plane deformation (Mode III or SH problem) to analyze dynamic ruptures propagating along a planar fault [11]. They derived the analytical expression of the non-hypersingular integration kernel with the piecewise constant discretization and successfully solved time-marching problems. The non-hypersingular BIEs for the 2-D in-plane deformation (Modes I and II or P-SV problem) were derived by Tada and Yamashita [32] including the off-fault plane field, allowing the analyses of non-planar faults. Tada and Madariaga [31] expressly presented the functional forms of the analytically obtained 2-D kernels under the piecewise constant discretization for the all components of the dislocation and the stress. The 3-D dynamic rupture problem was first concerned by Fukuyama and Madariaga [12] for a planar fault with a rectangular boundary element, and Aochi and Fukuyama [8] extended to non-planar fault geometry. Tada [29] derived a compact forms of the kernels for the rectangular elements, and the triangular elements was by Tada [30].

In seismology, half-space problems are important but in general, such analytic expressions of the kernels are uneasy to obtain. The mirror source approximation is used under 3-D lateral slip fault cases [8]. For vertical slip cases, Chen and Zhang [10] applied numerical integrations to compute the half-space version of the kernel expressed by complex integrations in a spectral domain. Ando and Okuyama [4] took a different approach by adopting virtual boundary elements obeying the free surface condition by using the analytic expression of the kernel in a full space. Kame and Kusakabe [17] extended the similar idea to semi-infinite media with different elastic properties contacted at boundaries.

The reduction of computational costs is also concerned for the phenomena of large temporal scales. The major problem of dynamic BIEM is that the computational costs depend on time steps N at more than $O(N)$ because of the dense matrix nature of the integration kernels. Aside from the well-known methods such as the fast multipole method [24], H-matrix [13, 26], or the classic fast Fourier transformation, a different concept of the technique is used in seismology, which includes the temporal truncation of the history integral for the spectral domain BIEM in 2-D [21] and in 3-D [20], and FDPM in the real space domain BIEM in 2-D [5] and in 3-D with the model size of 10^5 elements [2]. The similar idea is proposed in engineering [34] concerning a simple case of the delta function response of the displacement field. Additionally, Ando [2] proposed to hybrid FDPM with H-matrix enabling reduction of the dependence on the number of elements M. Due to temporal efficiency of FDPM or the truncation, they have the utility in seamless analyses of phenomena governed by very different time scales, which are important in seismology.

2 Non-hypersingular Spatiotemporal Boundary Integral Equations

Before detailing FDPM, it may be useful to outline the derivations of numerically applicable or discretized BIEs to avoid unnecessary confusion from the mathematical appearance. I will briefly present the procedures by summarizing the works by Cochard and Madariaga [11], Tada and Yamashita [32], Tada and Madariaga [31], Tada [29] and Tada [30].

The boundary conditions on the fault planes are described as the mixed boundary condition [11] given by $\sum_l \sigma_{kl}(\boldsymbol{x}, t)\, n_l = f_k(\boldsymbol{x}, t)$ on the ruptured area Γ and $\Delta u_k(\boldsymbol{x}, t) = 0$ on the otherwise area, where σ_{kl} $(k \neq l)$ denotes the kl-component of the shear stress at the location $\boldsymbol{x}$ and the time t, $f(.)$ an arbitrary function determined by a friction law, Δu_k the slip or dislocation discontinuity across the fault n_l and the unit normal vector to Γ. The classic Betti representation theorem (see Sects. 2.3–2.5 of Aki and Richards [1]) gives the displacement at receivers inside the elastic medium due to slip at sources as

$$u_k(\boldsymbol{x}, t) = \int_\Gamma \int_{-\infty}^{t} \Delta u_p(\boldsymbol{\xi}, \tau)\, n_q(\boldsymbol{\xi}) \Sigma_{pq/k}(\boldsymbol{x}, t-\tau; \boldsymbol{\xi}, 0)\, d\tau d\boldsymbol{\xi}, \qquad (1)$$

(where $\Sigma_{pq/k} = \lambda\delta_{pq}\partial G_{kl}/\partial x_l + \mu(\partial G_{qk}/\partial x_p + \partial G_{pk}/\partial x_q)$ is the pq-component of the stress Green's function due to the unit impulsive body force acting in the kth direction, associated with the 3-D displacement Green's function given for the infinite homogeneous medium case as

$$\begin{aligned} G_{kl}(\boldsymbol{x}, t-\tau; \boldsymbol{\xi}, 0) = \left[\beta^2/(4\pi\mu)\right] \Big\{ &\left[(3\gamma_k\gamma_l - \delta_{kl})/r^3\right](t-\tau) \times \\ &\times \left[H(t-\tau-r/\alpha) - H(t-\tau-r/\beta)\right] \\ &+ \left(1/\alpha^2\right)(\gamma_k\gamma_l/r)\,\delta(t-\tau-r/\alpha) \\ &- (1/\beta^2)(\gamma_k\gamma_l - \delta_{kl})/r\,\delta(t-\tau-r/\beta) \Big\} \end{aligned} \qquad (2)$$

with $r \equiv \sqrt{(x_1-\xi_1)^2 + (x_2-\xi_2)^2 + (x_3-\xi_3)^2}$, $\gamma_k \equiv (x_k - \xi_k)/r$ and $H(.)$ the Heaviside step function, respectively. The first term on the right-hand side is called the near-field term in the phase between P and S wave fronts, while the last two terms represent the wave fronts, called the far-field terms. The 2-D counterpart for the plane strain (the SH line source) deformation is given by a scalar function as $G_{33} = H(t-\tau-r/\beta) / \left[2\pi\mu\sqrt{(t-\tau)^2 - r^2/\beta^2)}\right]$.

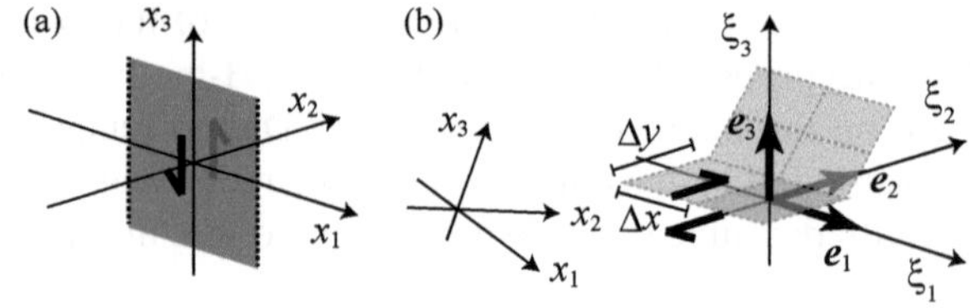

Fig. 1 Definition of the coordinate systems. **a** 2-D anti-plane case. **b** 3-D case for the global $x_1 - x_2 - x_3$ and the local $\xi_1 - \xi_2 - \xi_3$ coordinate systems defined associated with the centered boundary element colored dark gray

For the sake of a simple descriptive purpose, we summarize the following procedure in the simplest case of the on-fault field for the 2-D anti-plane strain case by following [11]; the fault plane is located on the $x_1 - x_3$-plane having infinite length in the x_3-direction, and the slip occurs in the x_3-direction (Fig. 1a). The boundary integral equation (BIE) representing the stress change is given by differentiating Eq. (1) with respect to $\boldsymbol{x}$ as $\sigma_{32}(x_1, t) = \mu \partial u_3 / \partial x_2$, and we obtain the expression of the kernel as $H[t - \tau - (x_1 - \xi_1)/\beta] / \left[(t-\tau)^2 - (x_1 - \xi_1)^2/\beta^2\right]^{3/2}$, which has the hypersingularity near the source, i.e., $\xi \to x$ and $\tau \to t$, and the wave front, $(t - \tau)^2 \to (x_3 - \xi_3)^2/\beta^2$. Note that the Heaviside function here describes the domain of influence, namely the causality cone. Next, to increase the numerical accuracy, the regularization of the hypersingular equation is performed by using the method proposed by Koller, Bonnet, and Madariaga [19] based on (i) the integration by parts to move the differentiations of the Green's function to those of the slip and (ii) transformation of the spatial differential to the temporal one engaging the form of the wave equation. Finally, we obtain

$$\begin{aligned}
&\sigma_{32}(x_1, t) \\
&= \frac{\mu}{2\pi} \int_\Gamma \frac{1}{x_1 - \xi_1} \int_{-\infty}^{t} \frac{(t-\tau) H[t-\tau-(x_1-\xi_1)/\beta]}{\left[(t-\tau)^2-(x_1-\xi_1)^2/\beta^2\right]^{1/2}} \frac{\partial}{\partial \xi} \Delta \dot{u}_3(\xi_1, \tau)\, d\tau d\xi_1 \\
&\quad - \frac{\mu}{2\pi\beta^2} \int_\Gamma \int_{-\infty}^{t} \frac{H[t-\tau-(x_1-\xi_1)/\beta]}{\left[(t-\tau)^2-(x_1-\xi_1)^2/\beta^2\right]^{1/2}} \frac{\partial}{\partial \tau} \Delta \dot{u}_3(\xi_1, \tau)\, d\tau d\xi_1 .
\end{aligned} \tag{3}$$

Discretization

Next, we aim to discretize Eq. (3) to be applicable to numerical analysis. Since the analytical expression of integration kernels reduces significantly the numerical cost, efforts have been made to integrate Eq. (3) with the appropriately discretized slip rate profiles $\Delta \dot{u}$. The often used is the piecewise constant interpolation of the slip rate, where the slip rate is approximated to be constant in each boundary element. There, the slip rate profile is approximated by the summation of the boxcar function associated with the collocation points $\left(x^j, t^m\right)$ over space and time given for the planar 2-D fault case by $\Delta \dot{u}_3(x_1, t) = \sum_j \sum_m D_3^{j,m}$ $\left[H\left(x_1 - x_1^j + \Delta x/2\right) - H\left(x_1 - x_1^j - \Delta x/2\right)\right]\left[H\left(t - t^m + \Delta t\right) - H\left(t - t^m\right)\right]$, where Δx and Δt are the spatial and temporal lengths of the boundary element, respectively. $D_3^{j,m}$ is the amplitude of the slip rate at the jth element and the mth time step. The superscripts denote the numbers in the discretization unless otherwise noted. There are parallel expressions for the cases of the 3-D rectangular [29] and triangular [30] elements and easily extended to the non-planar fault geometry cases by the combination of the planar elements.

The analytic integration of Eq. (3) substituting the above discretize slip rate is straightforward by noticing the relation $\partial H(s)/\partial s = \delta(s)$, and the discretized version of BIEs is shown as

$$\sigma_{pq}^{i,n} = -(\mu/2\beta)\left(D_k^{i,n} + \sum_j \sum_{m=0}^{n-1} \sum_{k=1}^{3} K_{pq/k}^{i,j,n-m} D_k^{j,m}\right), \tag{4}$$

where $K_{pq/k}^{i,j,n-m}$ is the integration kernel describing the pq-component of the stress at the ith element and nth time step as the response to the kth component of the slip rate D at the jth element and mth time step (a general expression for the components is presented here for the later use). The first term on the right-hand side is the radiation damping (or the instantaneous) term describing the instantaneous stress change due to the slip at the same element possible to be separated from the otherwise integral term when the Courant-Friedrichs-Lewy Condition is met for the values of Δx and Δt [31]. The form of the kernel K depends on the dimension of the problem and the shape of the elements. For example in the 2-D, Tada and Madariaga [31] obtained the expression including the off-plane field ($\xi_2 \neq 0$) necessary for the non-planar fault which is given as $K_{32/3}^{i,j,n-m} = I_{32/3}(\xi_L, \xi_2, t_S) - I_{32/3}(\xi_R, \xi_2, t_S) - I_{32/3}[\xi_L, \xi_2, t_L] + I_{32/3}[\xi_R, \xi_2, t_L]$ with $\xi_L = \xi_1^i - \xi_1^j + \Delta x/2$, $\xi_R = \xi_1^i - \xi_1^j - \Delta x/2$, $t_S = t^n - t^m + \Delta t$, $t_L = t^n - t^m$; the concrete expression is given as $I_{32/3}(\xi_1, \xi_2, t) = H(\xi_1) H(t - |\xi_2|/\beta) + (1/\pi)\,\mathrm{sgn}(\xi_1) H(t - r/\beta) \quad \left\{(|\xi_1|/r) \times \sqrt{(\beta t/r)^2 - 1} - \mathrm{Arccos}\left[|\xi_1|/\left(\beta U_\beta\right)\right]\right\}$ with $r = \sqrt{\xi_1^2 + \xi_2^2}$ and $U_\beta = \sqrt{t^2 - (\xi_2/\beta)^2}$. Each of $I(.)$ represents the response to the Heaviside step function, and the summation gives that to the boxcar function. Figure 1b schematically illustrated the treatment of the non-planar geometry where the location of a receiver defined in the $x_1 - x_2 - x_3$ global coordinate system is rotationally translated to that in the $\xi_1 - \xi_2 - \xi_3$ local coordinate system associated with a certain elements as $\xi_k = \boldsymbol{e}_k \cdot x$, with the basis vector $\boldsymbol{e}_k$. The concrete expressions for the 2-D stress fields are given as Eqs. (A1)–(A9) of Tada and Madariaga [31].

Extension to 3-D

The 3-D cases for the rectangular element laying on $x_1 - x_2$-plane with the size of $\Delta x \times \Delta y$ (see Fig. 1b) can be considered as $K_{pq/k}^{i,j,n-m} = L_{pq/k}\left(x_1^i - x_1^j, x_2^i - x_2^j, x_3^i - x_3^j, t_S\right) - L_{pq/k}(x_1^i - x_1^j, x_2^i - x_2^j, x_3^i - x_3^j, t_L)$ with $L_{pq/k}(\xi_1, \xi_2, \xi_3, t) = L_{pq/k}^o(\xi_{1L}, \xi_{2L}, \xi_3, t) - L_{pq/k}^o(\xi_{1R}, \xi_{2L}, \xi_3, t) - L_{pq/k}^o(\xi_{1L}, \xi_{2R}, \xi_3, t) + L_{pq/k}^o(\xi_{1R}, \xi_{2R}, \xi_3, t)$, where $\xi_{1L} = \xi_1 + \Delta x/2$, $\xi_{2L} = \xi_2 + \Delta y/2$, $\xi_{1R} = \xi_1 - \Delta x/2$, and $\xi_{2R} = \xi_2 - \Delta y/2$. The concrete expressions for $L_{pq/k}^o$, which corresponds to the elastic response to the quadrant-shaped semi-infinite source, are given by Eqs. (77) – (98) in Tada [29], where a typical example is shown as

$$
\begin{aligned}
L^{o}_{31/1} = &-\frac{2\beta^3\xi_1\left(3\xi_3^2-\xi_1^2\right)}{3\pi\chi_1^6}\left[g_1\left(\xi_1,\xi_2,\alpha\right)-g_1\left(\xi_1,\xi_2,\beta\right)\right] \\
&-\frac{\beta^3\xi_1\xi_2\xi_3^2}{\pi R^5\chi_1^2}\left[g_4\left(R,\alpha\right)-g_4\left(R,\beta\right)\right] \\
&-\frac{\beta\xi_1}{\pi\chi_1^2}\left\{\frac{2\beta^2\xi_3^2}{\alpha^2\chi_1^2}g_2\left(\xi_1,\xi_2,\alpha\right)\right. \\
&\left.-\left\{\frac{\xi_2\chi_1^2}{2\xi_1\left(\xi_2^2+\xi_3^2\right)}+\frac{2\xi_3^2}{\chi_1^2}\right\}g_2\left(\xi_2,\xi_1,\beta\right)\right\} \\
&-1/\left(2\pi\right)\left[g_3\left(\xi_1,\xi_2,\beta\right)+g_3\left(\xi_2,\xi_1,\beta\right)\right] \\
&+H\left(\xi_1\right)H\left(\xi_2\right)H\left(t-\left|\xi_3\right|/\beta\right)
\end{aligned}
\tag{5}
$$

with $R \equiv \sqrt{\xi_1^2+\xi_2^2+\xi_3^2}$ and $\chi_1 = \sqrt{\xi_1^2+\xi_3^2}$. The above equation involves the time dependencies only in the form classified as

$$
\left\{
\begin{aligned}
g_1\left(\xi_1,\xi_2,c\right) &\equiv \left[t^2-\chi_1^2/c^2\right]^{\frac{3}{2}}\left[2H\left(\xi_2\right)H\left(\tau_\chi\right)-\mathrm{sgn}\left(\xi_2\right)H\left(\tau_R\right)\right] \\
&\quad+\left\{\left[\xi_2\left(2R^2+\chi_1^2\right)/2R^3\right]t^3-\left[\left(3\xi_2\chi_1^2\right)/2c^2R\right]t\right\}H\left(\tau_R\right) \\
g_2\left(\xi_1,\xi_2,c\right) &\equiv \sqrt{t^2-\chi_1^2/c^2}\left[2H\left(\xi_2\right)H\left(\tau_\chi\right)-\mathrm{sgn}\left(\xi_2\right)H\left(\tau_R\right)\right]+\frac{\xi_2}{R}tH\left(\tau_R\right) \\
g_3\left(\xi_1,\xi_2,c\right) &\equiv \arctan\sqrt{t^2-\chi_1^2/c^2}\left[2H\left(\xi_2\right)H\left(\tau_\chi\right)-\mathrm{sgn}\left(\xi_2\right)H\left(\tau_R\right)\right] \\
&\quad+\left(\arctan\xi_2\right)/\xi_1 H\left(\tau_R\right) \\
g_4\left(R,c\right) &\equiv \left[t^3-\left(R^2/c^2\right)t\right]H\left(\tau_R\right) \\
g_5\left(R,c\right) &\equiv \left[t-R/c\right]H\left(\tau_R\right),
\end{aligned}
\right.
\tag{6}
$$

with the elastic wave speeds $c(=\alpha$ or $\beta)$, and $\tau_\chi = t - \chi_1/c$ and $\tau_R = t - R/c$; the Heaviside functions simply partition the domains in the causality cone (Fig. 2) used in FDPM. The triangular element case is similarly given by Eqs. (91)–(108) in Tada [30].

3 Fast Domain Partitioning Method (FDPM) and Quasi-dynamic Treatment

We are now ready to introduce FDPM, which is the technique that makes use of the particular functional forms representing physical characteristics of the spatiotemporal Green's function. This method makes it possible to reduce the CPU time and the memory requirement of the time-dependent problems, respectively, to $O(MN^2)$ and $O(M^2)$ from the original costs of $O(MN^3)$ and $O(M^2N)$. Such reduction is attained without degrading accuracy in 3-D. In this section, we will overview the basic concept with the emphasis on the suitability to seamless connection of the fully dynamic and quasi-dynamic simulation schemes. We derive the expressions in the rectangular elements below and the triangular elements in Ando [2]

The spatiotemporal stress kernel as in Eq. (5) consists of the domains physically describing the phases of the P and S wave fronts, the transient phase in-between these fronts, and the static equilibrium after the passage of S-front (see Fig. 2). We

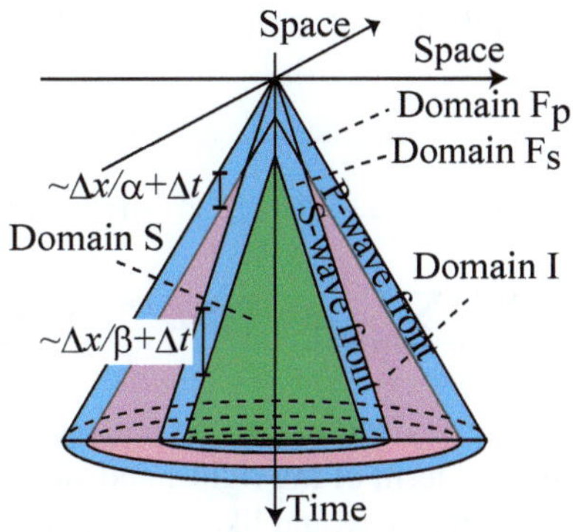

Fig. 2 Schematic illustration of the domains in the causality cone (the area of influence)

call these domains, respectively, as Domain F, Domain I, and Domain S, which have particular time dependences (Eq. (6)).

The main idea of FDPM is to decompose the original kernel matrix A of M elements in space and N steps in time to the product of the vectors like as

$$\begin{pmatrix} A_{11} & \cdots & A_{1N} \\ \vdots & \ddots & \vdots \\ A_{M1} & \cdots & A_{MN} \end{pmatrix} \to \begin{pmatrix} a_1 \\ \vdots \\ a_M \end{pmatrix} \left(b_1 \cdots b_N \right). \tag{7}$$

Due to the dense matrix nature of the integration kernels, a naive implementation (left-hand side) leads to the memory storage and the floating point computation of $M \times N$; however, they can be reduced to $M + N$ on the right-hand side for a certain condition. This decomposition is possible in Domains I and S.

Domain I

In Domain I, the functions g_n have the following reduced forms:

$$\begin{cases} g_1^I(\xi_1, \xi_2, \alpha) \equiv \left[\xi_2\left(2R^2 + \chi_1^2\right) / \left(2R^3\right)\right] t^3 - 3\xi_2\chi_1^2/(2\alpha^2 R)t \\ g_2^I(\xi_1, \xi_2, \alpha) \equiv (\xi_2/R)\, t \\ g_3^I(\xi_1, \xi_2, \alpha) \equiv \arctan\xi_2/\xi_1 \, (\to 0) \\ g_4^I(R, \alpha) \equiv t^3 - \left(R^2/\beta^2\right) t \\ g_5^I(R, \alpha) \equiv t - R/\alpha \, (\to t), \end{cases} \tag{8}$$

with $g_n^I(\xi_1, \xi_2, \beta) = 0$. Substitution of Eq. (8) to (5) yields

$$\begin{aligned} L_{31/1}^I \equiv & - \left\{ \frac{2\beta^3\xi_1\xi_2\left(3\xi_3^2 - \xi_1^2\right)\left(2R^2 + \chi_1^2\right)}{6\pi\chi_1^6 R^3} + \frac{\beta^3}{\pi R^5} \frac{\xi_1\xi_2\xi_3^2}{\chi_1^2} \right\} t^3 \\ & + \left[\frac{\beta^3\xi_1\xi_2\left(3\xi_3^2 - \xi_1^2\right)}{\pi\alpha^2\chi_1^4 R} + \frac{\beta}{\pi R^3} \frac{\xi_1\xi_2\xi_3^2}{\chi_1^2} - \frac{2\beta^3\xi_1\xi_2\xi_3^2}{\pi\alpha^2\chi_1^4 R} \right] t. \end{aligned} \tag{9}$$

This equation is a low-order polynomial on time t separated from the space ξ, which we can apply the decomposition as in Eq. (7). By applying the temporal differentiation, the kernel in this Domain K_I is obtained as

$$K_I^{i,j,n-m} \rightarrow K_{IA}^{i,j} G^{n-m} + K_{IB}^{i,j} \Delta t \tag{10}$$

with $G^{n-m} = (t^n - t^m)^3 - (t^n - t^m - \Delta t)^3$; therefore, the BIE is reduced as

$$\sum_j \sum_{m=0}^{n-1} K^{i,j,n-m} D^{j,m} \rightarrow \sum_j K_{IA}^{i,j} \sum_{m=0}^{n-1} G^{n-m} D^{j,m} + \sum_j K_{IB}^{i,j} S^j, \tag{11}$$

where S^j denotes the slip obtained by considering $\sum_m \Delta t D^{j,m} = S^j$. Therefore, the second term of the right-hand side represents the purely static influence.

Here, we would like to pay attention to the form of the first term on the right-hand side; besides the reduction of the required memory storage, this form has an advantage in the computation time even though the times of the multiplications are not actually reduced. This advantage comes from the efficiency of the data access on the memory due to the huge difference in the memory space size between $K^{i,j,n-m}$ of $O(M^2N)$ and G^{n-m} of $O(N)$. We have confirmed on a supercomputer environment that the bottleneck of the algorithm is on the data access associated with the convolution, so that the increasing the sequentiality of the data access greatly contributes to increase the efficiency of the memory access and the use of caches.

Domain S

Likewise, in Domain S, since the terms with t^3 raised by P wave are vanished with the arrival of S wave, the relevant terms of the functions g_n only retain t as

$$\begin{cases} g_1^S(\xi_1, \xi_2, c) \equiv -\left[3\xi_2\left(\xi_1^2 + \xi_3^2\right) / \left(2c^2 R\right)\right] t \\ g_2^S(\xi_1, \xi_2, c) \equiv (\xi_2/R) t \\ g_3^S(\xi_1, \xi_2, c) \equiv \arctan(\xi_2/\xi_1) \, (\rightarrow 0) \\ g_4^S(R, c) \equiv -\left(R^2/c^2\right) t \\ g_5^S(R, c) \equiv t - R/c \, (\rightarrow t), \end{cases} \tag{12}$$

and the BIEs are reduced as $\sum_j \sum_{m=0}^{n-1} K^{i,j,n-m} D^{j,m} \rightarrow \sum_j K_S^{i,j} S^j$, where K_s denotes the static kernel in Domain S, having the negligibly small numerical costs.

Domain F

Domain F is shaped like surficial bands covering the volume of the causality cone for both P and S waves (Fig. 2), in which volume is scaled by the time steps N. In fact, the time dependence is not reduced in this domain because of the necessity to describe the spatiotemporal distance from the propagating wave fronts. However, it provides a huge numerical merit that the band thicknesses are constant over time, determined by the finiteness of the boxcar function used to discretize the slip rate profiles. Namely, with counting the dependence also on M, the integration over this domain has the order of the surface $O(MN^2)$ reduced from the volume of the original BIEM $O(MN^3)$, and the computation speed is limited at $O(MN^2)$, as well as the

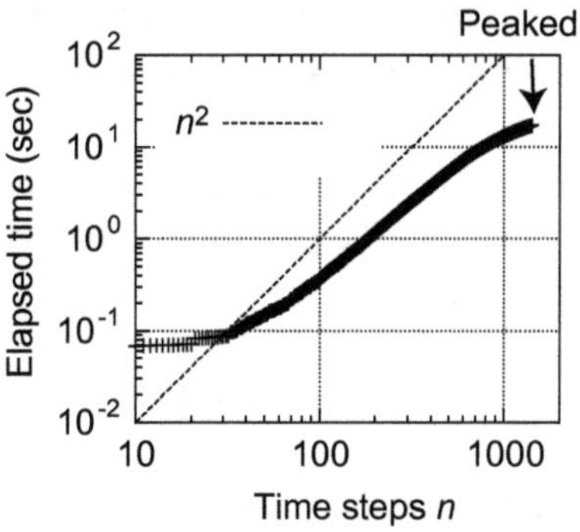

Fig. 3 Measured elapsed CPU time taken for the computation of each time step (modified after Ando [2])

required memory depending only on the number of the combinations for the sources and receivers as $O(M^2)$.

Figure 3 shows an example of the elapsed CPU time taken during each time step for a simulation in the case where the entire fault area starts to slip under the instantaneous stress reduction assumed as the initial condition. We can confirm that the elapsed time increases as the square of the time step n as we expected.

Relationship between FDPM and quasi-dynamic approaches

Quasi-dynamic approximation is intensively used in seismology to simulate effectively earthquake cycles, which consist of the recurrence of slow loading processes due to tectonic plate motions and rapid earthquake events. This approach invokes the static BIEs adding the instantaneous terms as $\sigma^{i,n} = -(\mu/2\beta)\, D^{i,n} + \sum_j K_S^{i,j} S^j$. On the other hand, FDPM concerns the fully dynamic BIEs as $\sigma^{i,n} = -(\mu/2\beta)\, D^{j,m} + \sum_j K^{i,j} \sum_{m=0}^{n-1} G^{n-m} D^{j,m} + \sum_j K_S^{i,j} S^j$. By comparing these equations, we can understand that quasi-dynamic approximation consists with negating the non-instantaneous time-dependent terms in the BIEs to avoid the time-consuming temporal integrations. This approximations may have validity if the inertia effects are small due to the slip is slow enough or the rapid slip events occurred in distant past, but it leads to a considerable error for an ongoing rapid slip event. As we recall Fig. 3, we can observe the CPU time tends to peak out as increasing time steps. This is because the integration domain of the BIE becomes dominated by efficiently evaluated Domain S; therefore, it is easily understood that the CPU time starts to decrease as the time marches further filled due to filled by Domain S. This procedure demonstrate how FDPM can seamlessly connect the quasi-static process and fully dynamic process. A simple implementation of an adaptive time step solvers is possible for FDPM to make the evaluation of Domain S further efficient like as other studies (e.g., [21]).

4 Numerical Implementation for Propagating Ruptures Obeying Frictions

In this section, I will provide a slightly technical description on numerical implementation. The governing equation is given by simultaneous equations constituted by BIEs

$$T_p^{i,n} = -\mu/(2\beta)\left(D_p^{i,n} + e_{3:\mathrm{a}}^i e_{p:b}^i \sum_j \sum_{m=0}^{n-1} \sum_k K_{ab/k}^{i,j,n-m} D_k^{j,m}\right) \tag{13}$$

and the boundary conditions as

$$\begin{cases} D_p^{i,n} = 0, & \text{if not fractured} \\ T_p^{i,n} = T_{resid:p}^{i,n} - T_{init:p}^{i}, & \text{if fractured} \end{cases}, \tag{14}$$

where $T_p^i \left(= e_{3:a}^i e_{p:b}^i \sigma_{ab}^i\right)$, $T_{init:p}^i$ and $T_{resid:p}^{i,n}$ denote, respectively, the pth component of the vectors of the traction change, and the initial and the residual tractions acting in the direction $\boldsymbol{e}_p^i$ $(p = 1 \text{ or } 2)$ on the ith element having the normal vector of $\boldsymbol{e}_3^i$. Our problem is to evaluate the shear traction $T_p^{i,n}$ around the crack tip and to obtain the slip rate $D^{i,n}$ at each time step under the mixed boundary condition given in Eq. (13). In seismological applications, faults are subjected to a compressional stress regime and the residual traction is described by friction. Note that shear dislocation in one direction on a planar element does not instantaneously change the normal traction, as well as the shear traction acting in the perpendicular direction. Therefore, the current piecewise planar discretization allows us to avoid the instantaneous effect of the normal traction on the friction, meaning $T_{resid}^{i,n} = \left(\mu_{fric}^{i,n} T_3^{i,n-1} + \sigma_o^i\right)$, where μ_{fric} and σ_o denote, respectively, the frictional coefficient and a constant cohesive strength.

The widely used friction law for the dynamic earthquake rupture simulation is called the linear slip-weakening law [7, 15], which can reproduce seismological observations (e.g., [16]) and the first-order characteristics of rock experiments referred as the rate- and state-dependent friction law (e.g., [27]). The frictional strength is decreased as the function of the amount of the slip $\left|\boldsymbol{S}^{i,n}\right| \left(= \left|\boldsymbol{D}^{i,n} \Delta t + \boldsymbol{S}^{i,n-1}\right|\right)$ with the characteristic slip distance D_c as $\mu_{firc}^{i,n} = \mu_{stat}^i - \left(\mu_{stat}^i - \mu_{dyn}^i\right) \left|\boldsymbol{S}^{i,n}\right| / D_c^i$, if $\left|\boldsymbol{S}^{i,n}\right| \le D_c^i$ or $= \mu_{dyn}^i$, if $\left|\boldsymbol{S}^{i,n}\right| \ge D_c^i$; by considering this equation, Eqs. (13) and (14) become a closed simultaneous equation with the two unknowns $D_1^{i,n}$ and $D_2^{i,n}$, which can be solved explicitly. Additionally, the fracture criterion is given by the condition $T_p^{i,n} + T_{init:p}^i \ge \mu^{i,n} T_3^{i,n-1} + \sigma_o$ outside of the area Γ be ruptured at the next time step.

5 Observational Constraints on the Fault Frictions and Crustal Stress States

We finally discuss how to constrain a physical parameter in nature. The average amount of the stress drop, which is the difference between the initial shear traction and the dynamic frictional strength as in Eq. (14), has been widely known to exhibit a quite similar value in the order of 1–10 M Pa for earthquakes regardless of the magnitudes (e.g., [18]). On the other hand, the estimate of the aforementioned characteristic slip distance D_c defining the fracture energy or toughness has difficulty in the seismological observations but some estimates show the typical amount of about 1 m (e.g., [16]).

The absolute stress at the seismogenic depth in the earth's crust is one of the most enigmatic parameters for the earth science since we are very much limited in ability of measuring in situ stress levels by drilling to the depth down to tens of kilometer. However, some clues are obtained from different types of observations.

Regarding the in situ stress measurements, the deepest borehole observation, called the KTB scientific drilling, has been made down to 8 Km at Germany [9]; they found the increased absolute stress levels for the horizontal principal stresses as the linear function of the depth, not only the overburden pressure. Also used is the seismic stress tensor inversion based on the observational analysis of the earthquake focal mechanism solutions [23]. The orientations of earthquake fault planes and the slip are determined by analyzing the polarity of the observed ground particle waveforms. By assuming that the slip occurs in the direction of the maximum shear traction acting on each fault plane, we can obtain the information about the directions of the every principal stresses and a ratio of the stress differences between them. Useful compilation maps are published (e.g., [14, 22]).

6 Summary

This paper presents the overview of the recent development history of ST-BIEM and FDPM, and the related issues in the field of seismology. Technical aspects are described regarding the application to the problems of dynamic rupture propagations as well as earthquake cycles. Typical problems in dealing with earthquake simulations are also discussed.

Acknowledgements This study is supported in part by JSPS/MEXT KAKENHI Grant Numbers JP25800253 and JP26109007.

References

1. Aki, K., Richards, P.G.: Quantitative Seismology, 2nd edn. University Science Books (2002)
2. Ando, R.: Fast Domain Partitioning Method for dynamic boundary integral equations applicable to non-planar faults dipping in 3-D elastic half-space. Geophys. J. Int. **207**, 833–847 (2016)
3. Ando, R., Okuyama, S.: Deep roots of upper plate faults and earthquake generation illuminated by volcanism. Geophys. Res. Lett. **37** (2010)
4. Ando, R., Yamashita, T.: Effects of mesoscopic-scale fault structure on dynamic earthquake ruptures: dynamic formation of geometrical complexity of earthquake faults. J. Geophys. Res. **112** (2007). https://doi.org/10.1029/2006JB004612
5. Ando, R., Kame, N., Yamashita, T.: An efficient boundary integral equation method applicable to the analysis of non-planar fault dynamics. Earth Planets Space **59**, 363–373 (2007)
6. Ando, R., Shaw, B.E., Scholz, C.H.: Quantifying natural fault geometry: statistics of splay fault angles. Bull. Seismol. Soc. Am. **99**, 389–395 (2009)
7. Andrews, D.J.: Rupture velocity of plane strain shear cracks. J. Geophys. Res. **81**, 5679–5687 (1976)
8. Aochi, H., Fukuyama, E.: Three-dimensional nonplanar simulation of the 1992 landers earthquake. J. Geophys. Res. 2002**107** (1992). https://doi.org/10.1029/2000JB000061
9. Brudy, M., Zoback, M.D., Fuchs, K., Rummel, F., Baumgartner, J.: Estimation of the complete stress tensor to 8 km depth in the KTB scientific drill holes: Implications for crustal strength. J. Geophys. Res. **102**, 18453–18475 (1997)
10. Chen, X.F., Zhang, H.M.: Modelling rupture dynamics of a planar fault in 3-D half space by boundary integral equation method: an overview. Pure Appl. Geophys. **163**, 267–299 (2006)
11. Cochard, A., Madariaga, R.: Dynamic faulting under rate-dependent friction. Pure Appl. Geophys. **142**, 419–445 (1994)
12. Fukuyama, E., Madariaga, R.: Rupture dynamics of a planar fault in a 3D elastic medium: rate- and slip-weakening friction. Bull. Seismol. Soc. Am. **88**, 1–17 (1998)
13. Hackbusch, W.: A sparse matrix arithmetic based on H-matrices. part I: introduction to H-matrices. Computing **62**, 89–108 (1999)
14. Hardebeck, J.L., Michael, A.J.: Stress orientations at intermediate angles to the San Andreas Fault, California. J. Geophys. Res. **109** (2004)
15. Ida, Y.: Cohesive force across the tip of a longitudinal-shear crack and Griffith's specific surface energy. J. Geophys. Res. **77**, 3796–3805 (1972)
16. Ide, S., Takeo, M.: Determination of constitutive relations of fault slip based on seismic wave analysis. J. Geophys. Res. **102**, 27379–27391 (1997)
17. Kame, N., Kusakabe, T.: Proposal of extended boundary integral equation method for rupture dynamics interacting with medium interfaces. J. Appl. Mech. Trans. ASME **79** (2012)
18. Kanamori, H., Anderson, D.L.: Theoretical basis of some empirical relations in seismology. Bull. Seismol. Soc. Am. **65**, 1073–1095 (1975)
19. Koller, M.G., Bonnet, M., Madariaga, R.: Modeling of dynamic crack-propagation using time-domain boundary integral-equations. Wave Motion **16**, 339–366 (1992)
20. Lapusta, N., Liu, Y.: Three-dimensional boundary integral modeling of spontaneous earthquake sequences and a seismic slip. J. Geophys. Res. **114** (2009)
21. Lapusta, N., Rice, J.R., Ben-Zion, Y., Zheng, G.T.: Elastodynamic analysis for slow tectonic loading with spontaneous rupture episodes on faults with rate- and state-dependent friction. J. Geophys. Res. **105**, 23765–23789 (2000)
22. Matsumoto, S. et al.: Spatial heterogeneities in tectonic stress in Kyushu, Japan and their relation to a major shear zone. Earth Planets Space **67** (2015)
23. Michael, A.J.: Use of focal mechanisms to determine stress - a control study. J. Geophys. Res. **92**, 357–368 (1987)
24. Nishimura, N.: Fast multipole accelerated boundary integral equation methods. Appl. Mech. Rev. **55**, 299–324 (2002)

25. Noda, H., Nakatani, M., Hori, T.: A slower fault may produce a smaller preseismic moment rate: non-1/t(f) acceleration of moment rate during nucleation and dependency on the background slip rate. Geophys. Res. Lett. **40**, 4850–4854 (2013)
26. Ohtani, M., et al.: Fast computation of quasi-dynamic earthquake cycle simulation with hierarchical matrices. Procedia Comput. Sci. **4**, 1456–1465 (2011)
27. Scholz, C.H.: Earthquakes and friction laws. Nature **391**, 37–42 (1998)
28. Scholz, C.H.: The Mechanics of Earthquakes and Faulting, 2nd edn. Cambridge University Press, Cambridge (2002)
29. Tada, T.: Displacement and stress Green's functions for a constant slip-rate on a quadrantal fault. Geophys. J. Int. **162**, 1007–1023 (2005)
30. Tada, T.: Stress Green's functions for a constant slip rate on a triangular fault. Geophys. J. Int. **164**, 653–669 (2006)
31. Tada, T., Madariaga, R.: Dynamic modelling of the flat 2-D crack by a semi-analytic BIEM scheme. Int. J. Numer. Methods Eng. **50**, 227–251 (2001)
32. Tada, T., Yamashita, T.: Non-hypersingular boundary integral equations for two-dimensional non-planar crack analysis. Geophys. J. Int. **130**, 269–282 (1997)
33. Tse, S.T., Rice, J.R.: Crustal earthquake instability in relation to the depth variation of frictional slip properties. J. Geophys. Res. **91**, 9452–9472 (1986)
34. Yoshikawa, H., Nishimura, N.: A study on the reduction of the memory requirements and the computational time for elastodynamic TD-BIEM using the temporal linearity of the fundamental solution. Trans. Jpn. Soc. Comp. Met. Eng. **9** (2009) No. 16-091211

Integral Representation and Its Applications in Earthquake Mechanics: A Review

Shiro Hirano

Abstract In seismology, a faulting process as a source is linked with an elastic wavefield as an observable not only via a partial differential equation (PDE) but also via an integral equation. We conduct a review of these links and focus on the latter in terms of forward/inverse analyses of kinematic/dynamic modeling, which are investigated by many seismologists. Difficulties in the analyses are also mentioned: estimation and hyper-singularity of an integration kernel, determination of the number of parameters for modeling, and assumed dynamic friction on faults.

1 Introduction

Earthquake faulting processes are modeled under a framework of dynamic fracture mechanics, which relates rupture extension to displacement/stress perturbations. The faulting processes feature (1) extension of a shear crack under compressive states, (2) rupture velocity up to the speeds of Rayleigh waves, shear waves, or rarely, compressional waves, and (3) balancing with friction of shear traction acting on a slipping surface, etc. We here offer a brief overview of analyses of faulting processes in seismology. We especially focus on an integral representation of dynamic rupture, which is equivalent to a time-dependent double layer potential. The representation was first developed in the frequency domain [26] and subsequently rewritten in the time domain [9]. Finally, Keiichi Aki completed it in a book [3, 4]. As mentioned in this article, the representation has been applied to seismic wavefield simulations (Sect. 3.1), inverse analyses of fault slip evolution (Sect. 3.2), and earthquake rupture simulations (Sect. 3.4). It has also been applied for inverse analyses of co-seismic stress change on faults, although this has only been done in a few studies (Sect. 3.3).

S. Hirano (✉)
College of Science and Engineering, Ritsumeikan University,
1-1-1, Noji-Higashi, Kusatsu-city, Shiga 525-8577, Japan
e-mail: s-hrn@fc.ritsumei.ac.jp

P. van Meurs et al. (eds.), *Mathematical Analysis of Continuum Mechanics and Industrial Applications II*, Mathematics for Industry 30,
https://doi.org/10.1007/978-981-10-6283-4_9

2 Formulation

2.1 *In Terms of PDE*

We start from the equilibrium condition of an elastic body. Let the domain Ω $\left(\subset \mathbb{R}^3\right)$ be the elastic body (e.g., the Earth) including a bounded surface Γ $(\subset \Omega)$ that represents an earthquake fault. Just before an earthquake, we assume that the elastic body is under the equilibrium condition, represented as

$$\nabla \cdot \sigma_0(\boldsymbol{x}) = 0, \quad \boldsymbol{x} \in \Omega \tag{1}$$

where σ_0 is an initial stress tensor in Ω. In this state, σ cannot be correlated with displacement or strain of Ω's interior because the Earth had undergone continuous displacement and had no reference state of the displacement or strain. Thus, when Eq. (1) is satisfied, we set a reference for the displacement and strain. Next, let $u_n(\boldsymbol{x}, t)$ be the n-th component of the displacement vector $\boldsymbol{u}(\boldsymbol{x}, t)$ $\left(\in \mathbb{R}^3\right)$ at the position $\boldsymbol{x}$ $(\in \Omega \backslash \Gamma)$ and time $t (\geq 0)$. As noted above, we assume that the displacement vector field had zero initial conditions: $\boldsymbol{u} = \partial_t \boldsymbol{u} = 0$ for $t = 0$. As in Fig. 1a, The stress perturbation tensor σ in Ω and normal and tangential (i.e., shear) components of the traction perturbation $\boldsymbol{N}$ and $\boldsymbol{T}$, respectively, at $\boldsymbol{\xi}$ $(\in \Gamma)$ are defined as

$$\sigma(\boldsymbol{x}, t) = \frac{1}{2} \boldsymbol{C}(\boldsymbol{x}) \left\{ \nabla \boldsymbol{u}(\boldsymbol{x}, t) + (\nabla \boldsymbol{u}(\boldsymbol{x}, t))^T \right\}, \tag{2}$$

$$\boldsymbol{N}(\boldsymbol{\xi}, t) = \lim_{\boldsymbol{x} \to \boldsymbol{\xi}} \left\{ \boldsymbol{\nu}^{\Gamma}(\boldsymbol{\xi}) \cdot \sigma(\boldsymbol{x}, t)\, \boldsymbol{\nu}^{\Gamma}(\boldsymbol{\xi}) \right\} \boldsymbol{\nu}^{\Gamma}(\boldsymbol{\xi}), \tag{3}$$

$$\boldsymbol{T}(\boldsymbol{\xi}, t) = \lim_{\boldsymbol{x} \to \boldsymbol{\xi}} \left\{ \sigma(\boldsymbol{x}, t)\, \boldsymbol{\nu}^{\Gamma}(\boldsymbol{\xi}) - \boldsymbol{N}(\boldsymbol{x}, t) \right\}, \tag{4}$$

where $\boldsymbol{C} = \left(c_{ijkl}\right)$ is a fourth-order elasticity tensor, and $\boldsymbol{\nu}^{\Gamma}$ is a local unit normal vector to Γ. Note that $\boldsymbol{N}$ and $\boldsymbol{T}$ are independent of the direction of the approach of $\boldsymbol{x}$ to $\boldsymbol{\xi}$ because of continuity of traction for any surfaces embedded in elastic bodies. Then, our concern is the following PDE and initial/boundary conditions for the displacement and stress perturbations:

$$\begin{cases} \rho(\boldsymbol{x}) \partial_t^2 \boldsymbol{u}(\boldsymbol{x}, t) = \nabla \cdot \sigma(\boldsymbol{x}, t), & \boldsymbol{x} \in \Omega \backslash \Gamma,\ t \geq 0 \quad (5a) \\ \boldsymbol{u}(\boldsymbol{x}, 0) = \partial_t \boldsymbol{u}(\boldsymbol{x}, 0) = 0, & \boldsymbol{x} \in \Omega \backslash \Gamma, \quad (5b) \\ \sigma(\boldsymbol{x}, t)\, \boldsymbol{\nu}^{\partial \Omega}(\boldsymbol{x}) = 0, & \boldsymbol{x} \in \partial \Omega \quad (5c) \end{cases}$$

where $\boldsymbol{\nu}^{\partial \Omega}$ is a local unit normal vector to $\partial \Omega$. Moreover, one of the following boundary conditions on the fault Γ is required at $\boldsymbol{\xi} \in \Gamma$ and $t \geq 0$:

$$\begin{cases} \lim_{\varepsilon \to 0} \left[\boldsymbol{u}\left(\boldsymbol{\xi} + \varepsilon \boldsymbol{\nu}^{\Gamma}(\boldsymbol{\xi}), t\right) - \boldsymbol{u}\left(\boldsymbol{\xi} - \varepsilon \boldsymbol{\nu}^{\Gamma}(\boldsymbol{\xi}), t\right) \right] = [\boldsymbol{u}](\boldsymbol{\xi}, t), \quad (6a) \\ \boldsymbol{T}_0(\boldsymbol{\xi}) + \boldsymbol{T}(\boldsymbol{\xi}, t) = \boldsymbol{f}(\boldsymbol{\xi}, t), \quad (6b) \end{cases}$$

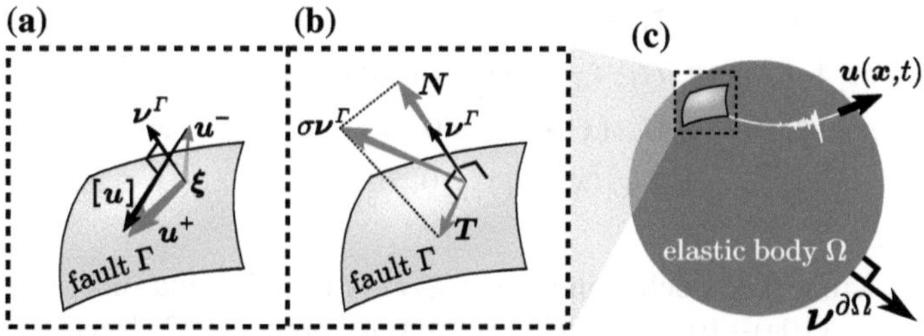

Fig. 1 Schematic illustrations of **a** slip $[\boldsymbol{u}] := \boldsymbol{u}^+ - \boldsymbol{u}^-$ parallel to the fault surface Γ embedded in the elastic body Ω, where $\boldsymbol{u}^+$ and $\boldsymbol{u}^-$ are displacement above and beyond $\boldsymbol{\xi} \in \Gamma$, respectively, and **b** the normal component $\boldsymbol{N}$ and tangential component $\boldsymbol{T}$ of the traction perturbation $\sigma\boldsymbol{\nu}^\Gamma$ due to the slip. **c** $\boldsymbol{u}(\boldsymbol{x}, t)$ is off-fault displacement due to the seismic wavefield

where $\boldsymbol{T}_0$, obtained by replacing σ in Eq. (4) by σ_0, is the background level of the tangential traction, and $\boldsymbol{f}$ is friction acting on the sliding surface Γ. In the right-hand side of (6a), $[\boldsymbol{u}]$ is the discontinuity of the displacement field (Fig. 1b). For earthquake faulting, we assume non-opening and non-penetration conditions represented as $[\boldsymbol{u}] \cdot \boldsymbol{\nu}^\Gamma = 0$ because the fault is under a highly compressive state; thus we regard $[\boldsymbol{u}]$ as slip. A boundary of the support of $[\boldsymbol{u}]$ is regarded as a rupture front, and its propagation velocity is the rupture velocity. In this regard, analyses of rupture propagation are included in analyses of $[\boldsymbol{u}]$.

2.2 *In Terms of Integral Equation*

In seismology, the essential idea is to relate the slip and displacement/stress perturbation at arbitrary points. The following equation is called the representation theorem of Aki and Richards [3, 4]:

$$\boldsymbol{u}(\boldsymbol{x}, t) = \int_\Gamma \boldsymbol{m} * \nabla_\xi \boldsymbol{G} \, d\boldsymbol{\xi} = \int_\Gamma \left\{ \int_0^t \boldsymbol{m}(\boldsymbol{\xi}, \tau) \nabla_\xi \boldsymbol{G}(\boldsymbol{x}, \boldsymbol{\xi}, t - \tau) \, d\tau \right\} d\boldsymbol{\xi}, \quad (7)$$

or, with summation convention,

$$u_n(\boldsymbol{x}, t) = \int_\Gamma m_{kl} * \partial_{\xi_l} G_{nk} \, d\boldsymbol{\xi}, \quad (8)$$

$$m_{kl}(\boldsymbol{\xi}, \tau) = [u_i](\boldsymbol{\xi}, \tau) \, \nu_j(\boldsymbol{\xi}) \, c_{ijkl}(\boldsymbol{\xi}) \quad (9)$$

where the asterisk $*$ denotes a convolution with respect to time from 0 to t, and the contracted tensor $\boldsymbol{m}(\boldsymbol{\xi}, t)$ defined in Eq. (9) is called the seismic moment density tensor or, hereinafter, moment tensor. In Eqs. (7) and (8), the second-order Green tensor $\boldsymbol{G}(\boldsymbol{x}, \boldsymbol{\xi}, t)$ is the solution to the following problem in Ω:

$$\begin{cases} \rho(\boldsymbol{x})\partial_t^2 G_{in}(\boldsymbol{x},\boldsymbol{\xi},t) \\ \quad = \partial_{x_j}\left\{c_{ijkl}(\boldsymbol{x})\partial_{x_l} G_{kn}(\boldsymbol{x},\boldsymbol{\xi},t)\right\} + \delta_{in}\delta(\boldsymbol{x}-\boldsymbol{\xi},t), & \boldsymbol{x}\in\Omega, t\geq 0, \quad (10a) \\ G_{nk}(\boldsymbol{x},\boldsymbol{\xi},0) = \partial_t G_{nk}(\boldsymbol{x},\boldsymbol{\xi},0) = 0, & \boldsymbol{x}\in\Omega\backslash\{\boldsymbol{\xi}\}, \quad (10b) \\ c_{ijkl}(\boldsymbol{x})\,\nu_j^{\partial\Omega}(\boldsymbol{x})\,\partial_{\xi_l} G_{kn}(\boldsymbol{x},\boldsymbol{\xi},t) = 0, & \boldsymbol{x}\in\partial\Omega, \quad (10c) \end{cases}$$

where δ_{in} is Kronecker's delta, and $\delta(\boldsymbol{x}-\boldsymbol{\xi},t)$ is Dirac's delta function. Note that $\boldsymbol{x}$ and $\boldsymbol{\xi}$ in Eqs. (10a)–(10c) are taken as arbitrary points in Ω. Hence, Eqs. (10a)–(10c) hold in an intact elastic body without the fault surface Γ. In other words, $\boldsymbol{G}$ is independent of Γ. Thanks to this nature, we can calculate $\boldsymbol{G}$ even if we do not know all faults embedded in the Earth.

Some advantages of the use of Eq. (7) in comparison with finite difference methods (FDMs) and finite element methods (FEMs) are the following. In Eq. (7), we consider quantities defined only on Γ to obtain the displacement at $\boldsymbol{x}$ while FDMs and FEMs require any information all over Ω. Hence, the integral representation requires very few numerically discretized elements. Moreover, when we consider traction on Γ on the basis of Eqs. (2)–(4), the boundary integral consists of the traction and moment tensor on Γ in an exact sense. On the other hand, in FDM or FEM, those quantities are sometimes not defined strictly along the fault surface but defined at different points each other in a staggered grid FDM or traditional FEM. Hence, use of Eq. (7) has an advantage in view of numerical accuracy and contributes to some analyses in seismology as follows.

3 Practical Problems

3.1 *Kinematic Forward Analysis of Seismic Wavefield:* $\boldsymbol{m} \mapsto \boldsymbol{u}$

Calculations of seismic wavefields are an important topic in terms of both geophysics and hazard assessment. Forward simulations of the wavefield are equivalent to solving Eqs. (5a)–(5c) with Eq. (6a). In many cases, these simulations have been performed using FDMs or FEMs with assumed spatio-temporal distributions of the moment tensor called scenario earthquakes, which are based on past earthquakes analyzed throughout the steps in Sect. 3.2, or on forward simulations shown in Sect. 3.4. By using seismometer networks, the near-surface seismic wavefield is directly observable in the aftermath of major earthquakes, allowing researchers to validate the results of their simulations by comparing calculated $\boldsymbol{u}(\boldsymbol{x},t)$ at $\boldsymbol{x}\in\partial\Omega$ and the observed waveforms. Because this procedure is independent of force, traction, stress, and friction, this is a kinematic modeling of the seismic wavefield.

In principle, the simulations can be executed using Eq. (7) if the kernel $\nabla_{\boldsymbol{\xi}}\boldsymbol{G}$ is available. In particular, reflection, refraction, scattering, and the intrinsic attenuation due to the heterogeneity of the Earth's interior are out of the scope for fundamental understanding on characteristics of seismic radiation from a faulting process. For this purpose, the kernel for an infinite homogeneous isotropic space has been employed. While the complete form of the kernel can be found in Eqs. (4.29)–(4.33) of Aki and Richards [4], the following approximation holds in the far field [3, 4, 11]:

$$\partial_{\xi_l} G_{nk} \sim \frac{\gamma_n \gamma_k \gamma_l}{4\pi\rho c_d^3} \frac{\partial_t \delta\left(t - \frac{r}{c_d}\right)}{r} - \left(\frac{\gamma_n \gamma_k - \delta_{nk}}{4\pi\rho c_s^3}\right) \gamma_l \frac{\partial_t \delta\left(t - \frac{r}{c_s}\right)}{r}, \tag{11}$$

where $\gamma_i = (x_i - \xi_i)/r, r = |\boldsymbol{x} - \boldsymbol{\xi}|$, and c_d and c_s are the speeds of dilatational (i.e., longitudinal or primary) and shear (i.e., transverse or secondary) waves, respectively. Using this kernel and the characteristics of a circular crack model, the spectra of far-field radiation have been studied analytically (e.g., [24, 31]) and numerically (e.g., [22, 24]). Moreover, Hok and Fukuyama [16] executed numerical calculations of ground motion considering a free surface, which is represented using the kernel for the infinite medium. Although the free surface was completely flat in their simulation, their method is applicable to any surface topography. Hence, their method will be applied to more scenario earthquakes for hazard assessment.

For more realistic simulations, a kernel that reflects the heterogeneity of the Earth is required. One powerful tool to obtain it is called an empirical Green's function method [20], as follows. Time-derivative of Eq. (7) results in

$$\partial_t \boldsymbol{u}(\boldsymbol{x}, t) = \int_\Gamma \partial_\tau \boldsymbol{m} * \nabla_{\boldsymbol{\xi}} \boldsymbol{G}\, d\boldsymbol{\xi}. \tag{12}$$

This means that, if a support of $\partial_\tau \boldsymbol{m}$ is sufficiently small in space and time (i.e., $\partial_\tau \boldsymbol{m}$ is pulse-like), we may assume that $\partial_\tau \boldsymbol{m} \sim \delta(\boldsymbol{\xi})\delta(\tau)$ holds. Therefore, an empirically observed ground velocity caused by a small earthquake on Γ, which can be actually observed by seismometers, approximates the kernel. Another tool to obtain the kernel is a straightforward calculation. Instead of solving Eqs. (10a)–(10c), a propagator matrix method has widely been employed. In this method, only radial and stratified heterogeneity of the Earth is assumed (i.e., $\boldsymbol{C}(\boldsymbol{x}) = \boldsymbol{C}(R)$, where R is the distance from the center of the Earth, and $\boldsymbol{C}$ is a step-wise function). Additionally, plane-wave decomposition, which is compatible with Snell's law, is considered; see the Chap. 9 in [4] or 6.5 in [32] for details. These two methods are also applicable to inverse analyses shown in Sect. 3.2.

3.2 *Kinematic Inverse Analysis of Fault Slip:* $\boldsymbol{u} \mapsto \boldsymbol{m}$

After major earthquakes occur, seismologists investigate what happened on a fault by analyzing seismograms, which record the ground velocity $\partial_t \boldsymbol{u}(\boldsymbol{x}, t)$ for $\boldsymbol{x} \in \partial\Omega$. Thus the time-derivative of Eq. (7) gives an inversion problem to obtain moment rate $\partial_\tau \boldsymbol{m}$ once $\partial_t \boldsymbol{u}^{\text{obs}}(\boldsymbol{x}, t)$ is observed. As with the previous subsection, this is a kinematic modeling of faulting. Effectively, the following equation is widely assumed:

$$\partial_t \boldsymbol{u}^{\mathrm{obs}}(\boldsymbol{x}, t) = \int_{\Gamma} \partial_\tau \boldsymbol{m} * \nabla_{\boldsymbol{\xi}} \boldsymbol{G}\, d\boldsymbol{\xi} + \boldsymbol{e}(\boldsymbol{x}, t), \tag{13}$$

where an error vector $\boldsymbol{e}$ is added because synthetic waveforms calculated via the integral on the right-hand side must not strictly agree with the observed ones owing to some errors in the observation system, modeling process, and/or uncertainty of the Green tensor. Discretized form of Eq. (13) is reduced to

$$\partial_t u_n^{\mathrm{obs}}\left(\boldsymbol{x}^i, t^j\right) = \partial_\tau m_{kl}\left(\boldsymbol{\xi}^p, \tau^q\right) \partial_{\xi_l} G_{kn}\left(\boldsymbol{x}^i, \boldsymbol{\xi}^p, t^j - \tau^q\right) \delta\boldsymbol{\xi}\, \delta\tau + e_n\left(\boldsymbol{x}^i, t^j\right), \tag{14}$$

where i and j are up to the numbers of observation points and samples of time series, respectively; and p and q are up to P and Q, the numbers of spatial and temporal discretization for the source process, respectively. Using linear equation solvers, we obtain some approximated solutions that minimize the norm of $\boldsymbol{e}$ (e.g., a (damped) least square solution [23]). However, P and Q are arbitrary, and the error reduces to an infinitesimal value as P and Q increase. Overfitted solutions in such a case are not robust, and we have to choose appropriate values of P and Q based on certain criteria. In many cases, Akaike's information criterion, which is equivalent to the principle of maximum entropy, is employed [1]. Moreover, the Bayesian information criterion has been employed to determine unknown hyper parameters that describe the smoothness of the solution on the basis of prior constraints from a physical point of view [2, 36]. More reviews on a practical procedure and results are found in [18], and a project to validate various schemes to solve this problem is on-going [25].

3.3 Inverse Analysis of Fault Dynamics: $\boldsymbol{m} \mapsto -\boldsymbol{T} = \boldsymbol{T}_0 - \boldsymbol{f}$

In seismology, traction and friction acting on fault surfaces are not directly observable. Given the inverted moment tensor $\boldsymbol{m}$, investigating the evolution of co-seismic traction $\boldsymbol{T}(\boldsymbol{x}, t)$ is a challenging problem. Substituting Eq. (7) to Eqs. (2)–(4), we obtain a linear relationship between them as Γ:

$$\boldsymbol{T}(\boldsymbol{x}, t) = \mathfrak{T} \int_{\Gamma} \boldsymbol{m} * \nabla_{\boldsymbol{\xi}} \boldsymbol{G}\, d\boldsymbol{\xi}, \tag{15}$$

where the linear operator $\mathfrak{T}$ maps any displacement field to shear traction perturbation on Γ resulting from the combination of Eqs. (2)–(4). This equation and a relation $-\boldsymbol{T}(\boldsymbol{x}, t) = \boldsymbol{T}_0(\boldsymbol{x}) - \boldsymbol{f}(\boldsymbol{x}, t)$ from Eq. (6b) enables the estimation of dynamic stress drop, temporal variation of the difference between initial traction $\boldsymbol{T}_0$ and friction $\boldsymbol{f}$ on the fault. In other words, it is not possible to estimate their magnitude separately using only seismic data.

In general, after seismic inversion analyses to obtain the moment tensor, $\boldsymbol{T}$ is inverted using an FDM [19, 27] or by assuming some characteristic form of the moment rate $\partial_\tau \boldsymbol{m}$ [35]. In principle, however, we can calculate the shear traction perturbation using Eq. (15). This calculation is, in fact, not straightforward because the integration kernel $\mathfrak{T}\nabla_{\xi}\boldsymbol{G}(\boldsymbol{x}, \boldsymbol{\xi}, t-\tau)$ has a hyper singularity for $\boldsymbol{x} \sim \boldsymbol{\xi}$ as shown in Sect. 4; note that the integration on the right-hand side of Eq. (15) is taken over Γ including the evaluation point $\boldsymbol{x}$ of the left-hand side. If Ω is an infinite homogeneous isotropic medium and Γ is flat, it is represented by Eq. (15) in the frequency-wave number domain according to Bouchon [8]:

$$\mathfrak{F}\boldsymbol{T}(\boldsymbol{k}, \omega) = \mathfrak{F}\boldsymbol{m}(\boldsymbol{k}, \omega)\mathfrak{F}\left(\mathfrak{T}\nabla_{\xi}\boldsymbol{G}\right)(\boldsymbol{k}, \omega), \tag{16}$$

where Fourier transform with respect to space and time

$$\mathfrak{F}f(\boldsymbol{k}, \omega) = \int_{\mathbf{R}^1} \left\{ \int_{\Gamma} f(\boldsymbol{x}, t) e^{i\boldsymbol{k}\cdot\boldsymbol{x}}\, d\boldsymbol{x} \right\} e^{-i\omega t}\, dt$$

is applied. However, the spectrum of the kernel could diverge in the wave number domain due to the hyper-singularity. Indeed, the amplitude of the kernel $\left|\mathfrak{F}\left(\mathfrak{T}\nabla_{\xi}\boldsymbol{G}\right)\right|$ is an increasing function of $|\boldsymbol{k}|$, such that an inverse Fourier transformation of (16) exists only when $\boldsymbol{m}$ is sufficiently smooth. The difficulty due to the hyper-singularity retained in the spectral domain is reduced by reguralizing the kernel in the time domain as discussed in Sect. 4.

3.4 *Forward Analysis of Fault Dynamics:* $-\boldsymbol{T} = \boldsymbol{T}_0 - \boldsymbol{f} \mapsto \boldsymbol{m}$

In mechanics, obtaining movements under given force and/or stress states is a forward problem. In our case, the fault slip is obtained after assuming the stress drop. This is quite a difficult problem because we cannot measure $\boldsymbol{T}_0$ or $\boldsymbol{f}$ independently and seismologically. $\boldsymbol{T}_0$ is considered to be exceedingly heterogeneous on faults because Earth's crust (i.e., surroundings of faults) has multi-scale structures. Moreover, in general, $\boldsymbol{f}$ non-linearly depends on the intrinsic state variable θ of the slipping surface [5, 10, 12, 28], normal stress $\boldsymbol{N}$, slip $[\boldsymbol{u}]$, and slip velocity $\partial_t[\boldsymbol{u}]$. Hence,

$$\boldsymbol{T}_0(\boldsymbol{x}) - \boldsymbol{f}(\theta, \boldsymbol{N}, [\boldsymbol{u}], \partial_t[\boldsymbol{u}]) = -\mathfrak{T}\int_{\Gamma} \boldsymbol{m} * \nabla_{\xi}\boldsymbol{G}\, d\boldsymbol{\xi} \tag{17}$$

holds. Equation (17) is a non-linear integro-differential equation of fault slip. To solve a discretized form of Eq. (17), the Boundary Integral Equation Method (BIEM, or Boundary Element Method, BEM), which is widely employed for physical modeling, is used; see Chapter "On Applications of Fast Domain Partitioning Method to Earthquake Simulations with Spatiotemporal Boundary Integral Equation Method" for its history and procedures. This is equivalent to solving Eqs. (5a)–(5c) with Eq. (6b).

In many fields, including acoustics and electromagnetics, such an integral is sometimes considered in the frequency domain. Analyses in the frequency domain are powerful tools when the focus is on phenomena considering a certain frequency. However, no specific frequency can be considered in fault mechanics. Instead, we consider fracture criteria that must hold at every moment, such that Eq. (17) can be solved in the time domain. Under the maximum stress criterion, $|\boldsymbol{f}| \leq f_s$ must hold at any given time, where f_s is the maximum static friction. Under the Griffith fracture criterion, the work rate $G = \int_{\Gamma} \boldsymbol{f} \cdot [\dot{\boldsymbol{u}}]\, d\boldsymbol{\xi}$ must be balanced with the fracture energy G_C at any given time.

Friction $\boldsymbol{f}$ has been investigated throughout laboratory experiments with various rocks. Recent reviews and advances on very-low slip velocity less than 10^{-2} m/s have been concentrated on the temporal evolution and physical background of friction [10, 12, 28]. In fact, co-seismic slip velocity is typically within the range of ~1–10 m/s, where no consensus on a unified friction law is reached. The slip-weakening law, which was originally introduced in a numerical modeling [17] and consequently confirmed in a laboratory [29], is widely employed in simulations:

$$\frac{|\boldsymbol{f}|}{|\boldsymbol{N}|} =(\mu_s - \mu_d)\phi\left(\frac{D}{D_c}\right) + \mu_d, \tag{18}$$

where μ_s and μ_d are the static and dynamic friction coefficients, respectively. $D = \|[\boldsymbol{u}]\|$ is slip distance, and $\phi(x)$ is a continuous and monotonically decreasing function from $\phi(0) = 1$ to $\phi(x)|_{1\leq x} = 0$. In many studies, $\phi(x) = (1 - x)H(1 - x)$, where $H(\cdot)$ is the Heaviside function, is employed for simplicity. The characteristic slip-weakening distance D_c is a critical parameter describing the law. In slip-weakening friction, the path-independent integral that should be balanced with the fracture energy is represented as follows [30]:

$$J - \mu_d\,|\boldsymbol{N}|\,D_c = \int_0^{D_c} (|\boldsymbol{f}| - \mu_d|\boldsymbol{N}|)\; dD,$$

where the left-hand side is path-independent. On the other hand, other friction laws for the co-seismic slip velocity range has been suggested. For example, a velocity- and state-dependent friction law based on thermodynamics on the slipping plane

$$\frac{|\boldsymbol{f}|}{|\boldsymbol{N}|} =\mu_s + \alpha\frac{V}{V + V_c} - \beta\frac{\theta}{\theta + V_c}, \tag{19}$$

$$\partial_t\theta =\frac{V - \theta}{D_c/V_c}, \tag{20}$$

has been employed [5], where $V = |\partial_t[\boldsymbol{u}]|$, and α, β, and V_c are positive constants. Using this law, an observed pulse-like rupture, where the duration of slip is significantly shorter than that of the entire faulting process [13], can be reproduced.

Table 1 Summary of forward/inverse analyses of fault kinematics/dynamics

		Forward	Inverse
Kinematics	Equation	(7)	(13)
	Based on	Assumed fault slip $[\boldsymbol{u}]$	Observed seismic wavefield $\boldsymbol{u}$
	Solving for	Seismic wavefield $\boldsymbol{u}$	Fault slip $[\boldsymbol{u}]$
Dynamics	Equation	(17)	(15)
	Based on	Assumed stress drop $-\boldsymbol{T}$	Estimated fault slip $[\boldsymbol{u}]$
	Solving for	Fault slip $[\boldsymbol{u}]$	Stress drop $-\boldsymbol{T}$

4 Summary and Discussion

As shown in the previous section, the representation theorem (7) reveals relationships among seismic wavefield, slip, and stress drop on the fault and enables their analysis; see summary in Table 1. In this section, we discuss a difficulty experienced during the execution of the analyses.

In reality, calculations using these equations are not archived literally because of the singularity of the integration kernel $\nabla_\xi \boldsymbol{G}$. An asymptotic representation of $\nabla_\xi \boldsymbol{G}$ is $\sim \partial_t \delta(t - r/v)/r$ for the far field (i.e., larger r) and $\sim \delta(t - r/v)/r^2$ for the near field (i.e., smaller r) [3, 4], where $r = |\boldsymbol{x} - \boldsymbol{\xi}|$ is distance between the source and the observation point, and v is the propagation speed of the wave.

In cases of the kinematic modeling in Sects. 3.1 and 3.2, the singularity does not matter. For Eqs. (7) and (13), the displacement field is observed at the far field in many cases. Therefore, the singularity of $\sim \partial_t \delta(t - r/v)/r$ dominates, and the denominator is non-singular in the far field. The numerator, Dirac's delta function, is unrealistic for modeling seismic waves because intrinsic attenuation of Earth's interior prevents the propagation of high-frequency components of waves towards the far field. Hence, we should consider the effective application of some low-pass filter $F(t)$ to the kernel [37]:

$$\partial_t \boldsymbol{u}^{\text{obs}} \sim \int_\Gamma \partial_\tau \boldsymbol{m} * \left(F * \nabla_\xi \boldsymbol{G}\right) d\boldsymbol{\xi} + \boldsymbol{e}.$$

As a result, the asymptotic form of the kernel $\sim \partial_t F(t - r/v)/r$ is finite and manageable in numerical schemes. Conventionally, F has been modeled as $|\mathfrak{F} F(\omega)| \sim e^{-\omega/Q}$ in the frequency domain, where the dimensionless parameter Q is the fractional energy loss per cycle; see Sect. 5.5 in [4] or 6.6 in [32]. Needless to say, the low-pass filtered kernel lacks shorter wavelength components. Thus, the resolution of the solution is reduced and depends on the characteristics of the attenuation. Effectively, the power of the filter is reduced by more than ~1 Hz in the frequency domain, which is equivalent to smaller than ~6 km in space because a typical dilatational wave speed is ~6 km/s in Earth's crust. The lack of resolution affects not only the fault slip inversion but also estimations of co-seismic potential energy release and kinetic energy radiation [14], which have been controversial in seismology. Given the empirical power spectral density of slip distributions and theoretical consideration, estimation of the energy may become more difficult than that of slip [14].

The situation is more severe in the mechanical modeling in Sects. 3.3 and 3.4 because the near-field (i.e., $\boldsymbol{x} \sim \boldsymbol{\xi}$) traction perturbation must be considered for Eqs. (15) and (17). Aggravating the situation, the operator $\mathfrak{T}$ consists of the first-order derivative, such that the singularity of $\mathfrak{T}\left(\nabla_{\xi} \boldsymbol{G}\right) \sim \delta(t-r/v)/r^3$ becomes dominant. The regularization of the hyper-singular integrals with integrations by parts is a method for solving this difficulty. For an infinite homogeneous isotropic medium (i.e., Ω corresponds to $\mathbb{R}^3$, and $C(\boldsymbol{x})$ is constant), Tada derived a regularized kernel of displacement and traction perturbation for a quadrantal fault [33] and a triangular fault [34]. In this manner, we can analyze the above integral equations with a numerically modeled fault surface that consists of polygons. Moreover, computational costs to treat Tada's regularized kernel has been dramatically reduced even with a free surface [6], [Chapter "On Applications of Fast Domain Partitioning Method to Earthquake Simulations with Spatiotemporal Boundary Integral Equation Method"]. Hence, the inverse modeling of co-seismic traction perturbation in Sect. 3.3 via Eq. (15) will be used more actively as in [16] while the forward modeling in Sect. 3.4 with non-planar faults have already been applied in a homogeneous full-space [7] and a half-space [6, 16]. In addition, applications of regularized integral equations to heterogeneous elastic 2-D media are developing (e.g., anti-plane faulting in two-welded half spaces [15], and piecewise-homogeneous medium [21]). Because surroundings of faults are generally heterogeneous, the focus of future researches will be on numerical schemes to treat interactions between such media and also dynamic faulting processes in the 3-D domain.

Acknowledgements The author is grateful to T. Yamaguchi and T. Kusakabe for helpful discussions. Comments from a reviewer contributed to improve the manuscript.

References

1. Akaike, H.: On entropy maximization principle. In: Krishnaiah, P.R. (ed.) Application of Statistics. pp. 27–41. North-Holland, Amsterdam (1977)
2. Akaike, H.: Likelihood and the Bayes procedure. In: Bernardo, J.M., DeGroot, M.H., Lindley, D.V., Smith, A.F.M. (eds.) Baysian Statistics, pp. 143–166. University Press, Valencia, Spain (1980)
3. Aki, K., Richards, P.G.: Quantitative seismology, 1st edn. Freeman (1980)
4. Aki, K., Richards, P.G.: Quantitative seismology, 2nd edn. University Science Books (2002)
5. Ampuero, J.-P., Ben-Zion, Y.: Cracks, pulses and macroscopic asymmetry of dynamic rupture on a bimaterial interface with velocity-weakening friction. Geophys. J. Int. **173**, 674–692 (2008)
6. Ando, R.: Fast domain partitioning method for dynamic boundary integral equations applicable to non-planar faults dipping in 3-D elastic half-space. Geophys. J. Int. **207**, 833–847 (2016)
7. Aochi, H., Fukuyama, E., Matsu'ura, M.: Spontaneous rupture propagation on a non-planar fault in 3-D elastic medium. Pure Appl. Geophys. **157**, 2003–2027 (2000)

8. Bouchon, M.: The state of stress on some faults of the San Andreas system as inferred from near-field strong motion data. J. Geophys. Res. **102**(B6), 11731–11744 (1997)
9. Burridge, R., Knopoff, L.: Body force equivalents for seismic dislocations. Bull. Seism. Soc. Am. **54**(6), 1875–1888 (1964)
10. Dieterich, J.H.: Applications of rate- and state-dependent friction to models of fault slip and earthquake occurrence. In: Kanamori, H. (ed.) Treatise on Geophysics. Earthquake Seismology, vol. 4, pp. 107–129. Elsevier, Amsterdam (2007)
11. Haskell, N.A.: Total energy and energy spectral density of elastic wave radiation from propagating faults. Bull. Seism. Soc. Am. **54**(6), 1811–1841 (1964)
12. Hatano, T.: Friction laws from dimensional-analysis point of view. Geophys. J. Int. **202**, 2159–2162 (2015)
13. Heaton, T.H.: Evidence for and implications of self-healing pulses of slip in earthquake rupture. Phys. Earth Planet. Int. **64**, 1–20 (1990)
14. Hirano, S., Yagi, Y.: Dependence of seismic and radiated energy on shorter wavelength components. Geophys. J. Int. **209**(3), 1585–1592
15. Hirano, S., Yamashita, T.: Dynamic antiplane rupture propagation crossing a material interface: modelling with BIEM. Geophys. J. Int. **200**(2), 1222–1235 (2015)
16. Hok, S., Fukuyama, E.: A new BIEM for rupture dynamics in half-space and its application to the 2008 Iwate-Miyagi Nairiku earthquake. Geophys. J. Int. **184**, 301–324 (2011)
17. Ida, Y.: Cohesive force across the tip of a longitudinal-shear crack and Griffith's specific surface energy. J. Geophys. Res. **77**(20), 3796–3805 (1972)
18. Ide, S., Slip inversion. In: Kanamori, H. (ed.) Treatise on Geophysics. Earthquake Seismology, vol. 4, pp. 193–224. Elsevier (2007)
19. Ide, S., Takeo, M.: Determination of constitutive relations of fault slip based on seismic wave analysis. J. Geophys. Res. B **102**(12), 27379–27391 (1997)
20. Irikura, K.: Prediction of strong acceleration motion using empirical Green's function. In: Proceedings of the 7th Japan Earthquake Engineering Symposium, pp. 151–156 (1986)
21. Kame, N., Kusakabe, T.: Proposal of extended boundary integral equation method for rupture dynamics interacting with medium interfaces. J. App. Mech. **79**(3), 031017 (2012)
22. Kaneko, Y., Shearer, P.M.: Seismic source spectra and estimated stress drop derived from cohesive-zone models of circular subshear rupture. Geophys. J. Int. **197**(2), 1002–1015 (2014)
23. Kikuchi, M., Kanamori, H.: Inversion of complex body waves-III. Bull. Seism. Soc. Am. **81**, 2335–2350 (1991)
24. Madariaga, R.: High-frequency radiation from crack (stress drop) models of earthquake faulting. Geophys. J. R. Astr. Soc. **51**, 625–651 (1977)
25. Mai, P.M., Schorlemmer, D., Page, M., Ampuero, J-P., Asano, K., Causse, M., Custodio, S., Fan, W., Festa, G., Galis, M., Gallovic, F., Imperatori, W., Kser, M., Malytskyy, D., Okuwaki, R., Pollitz, F., Passone, L., Razafindrakoto, H., Sekiguchi, H., Song, S., Somala, S., Thingbaijam, K., Twardzik, C., van Driel, M., Vyas, J., Wang, R., Yagi, Y., Zielke, O.: The earthquake-source inversion validation (SIV) project. Seism. Res. Lett. **87**(3), 690–708 (2016)
26. Maruyama, T.: On the force equivalents of dynamical elastic dislocations with reference to the earthquake mechanism. Bull. Earthq. Res. Inst. **41**, 467–486 (1963)
27. Mikumo, T., Olsen, K.B., Fukuyama, E., Yagi, Y.: Stress-breakdown time and slip-weakening distance inferred from slip-velocity functions on earthquake faults. Bull. Seism. Soc. Am. **93**(1), 264–282 (2003)
28. Nagata, K., Nakatani, M., Yoshida, S.: A revised rate- and state-dependent friction law obtained by constraining constitutive and evolution laws separately with laboratory data. J. Geophys. Res. B **117**, 02314 (2012)
29. Ohnaka, M., Yamashita, T.: A cohesive zone model for dynamic shear faulting based on experimentally inferred constitutive relation and strong motion source parameters. J. Geophys. Res. B **94**(4), 4089–4104 (1989)
30. Palmer, A.C., Rice, J.R.: The growth of slip surfaces in the progressive failure of over-consolidated clay. Proc. R. Soc. Lond. A. **332**, 527–548 (1973)

31. Sato, T., Hirasawa, T.: Body wave spectra from propagating shear cracks. J. Phys. Earth. **21**, 415–431 (1973)
32. Shearer, P.M.: Introduction to Seismology, 2nd edn. Cambridge university press, Cambridge (2009)
33. Tada, T.: Displacement and stress Green's functions for a constant slip-rate on a quadrantal fault. Geophys. J. Int. **162**, 1007–1023 (2005)
34. Tada, T.: Stress Green's functions for a constant slip rate on a triangular fault. Geophys. J. Int. **164**, 653–669 (2006)
35. Tinti, E., Fukuyama, E., Piatanesi, A., Cocco, M.: A kinematic source-time function compatible with earthquake dynamics. Bull. Seism. Soc. Am. **95**(4), 1211–1223 (2005)
36. Yabuki, T., Matsu'ura, M.: Geodetic data inversion using a Bayesian information criterion for spatial distribution of fault slip. Geophys. J. Int. **109**(2), 363–375 (1992)
37. Yagi, Y., Fukahata, Y.: Introduction of uncetainty of Green's function into waveform inversion for seismic source processes. Geophys. J. Int. **186**, 711–720 (2011)

Part IV
Material Structure

Brief Introduction to Damage Mechanics and Its Relation to Deformations

Simon Grützner and Adrian Muntean

Abstract We discuss some principle concepts of damage mechanics and outline a possibility to address the open question of the damage-to-deformation relation by suggesting a parameter identification setting. To this end, we introduce a variable motivated by the physical damage phenomenon and comment on its accessibility through measurements. We give an extensive survey on analytic results and present an isotropic irreversible partial damage model in a dynamic mechanical setting in form of a second-order hyperbolic equation coupled with an ordinary differential equation for the damage evolution. We end with a note on a possible parameter identification setting.

2010 Mathematics Subject Classification codes: 74-06 · 74A45 · 74B20 · 74G75

1 Introduction

Any real material is subjected to damage and its evolution which causes crucial mechanical properties such as stiffness, load-carrying capacity, fatigue limit, or creep rupture time to change. Contrary to today's growing demands especially on the mechanical properties of industrially used materials, most standards are based on elastic material behavior and, hence, neglect the effect damage has on them. This has led to a growing interest in damage mechanics as a general research topic for the last 20 years and to a never-ending search for more accurate models. Engineers have developed many sophisticated models during these years. While some of the research is directed toward the derivation within the framework of non-equilibrium thermodynamics and other focused on numeric investigations, experimental results are quite rare. One reason for this is that damage cannot be measured directly and

S. Grützner (✉)
ZeTeM - Center for Industrial Mathematics, University of Bremen, Bremen, Germany
e-mail: sg@math.uni-bremen.de

A. Muntean
Department of Mathematics and Computer Science, Karlstad University, Karlstad, Sweden
e-mail: adrian.muntean@kau.se

P. van Meurs et al. (eds.), *Mathematical Analysis of Continuum Mechanics and Industrial Applications II*, Mathematics for Industry 30,
https://doi.org/10.1007/978-981-10-6283-4_10

has to be determined through intermediate physical quantities it affects. Hence, measuring damage indirectly strongly depends on the assumed relation between damage and the observable quantity. Concerning this relation, however, there is no general consensus yet, which is the true motivation for our project.

Starting from the common understanding that damage affects the elastic response of a material and itself depends on the deformation state, the information gathered from a standard material test for determining elastic parameters is used to investigate how the test specimen's deformation contributes to the damage evolution. This open question can be addressed in a parameter identification setting. However, in this current work, we focus on presenting the direct problem in the shape of a parameter-to-data map and comment on some of its basic mathematical properties. To this end, this work is structured in the following way. We continue this section with a brief introduction into the concept of damage followed by a heuristic motivation for defining an isotropic damage variable. Since it is a crucial point in our long-term project to asses the damage state through experimental data, we comment on the accessibility of damage via measurements in Sect. 1.3. To relate our work to active research in the field of damage modeling and to emphasize the benefits of our model, we give an extensive survey on mathematical investigations of such models in Sect. 1.4 before introducing our model in Sect. 2. We conclude with a few comments concerning the well-posedness of our damage evolution model.

1.1 Concept of Damage

Usually already in its original state, a material contains a number of defects such as microvoids and micropores. During a deformation process, these inner voids grow and merge. In addition, inhomogenities present in these materials such as inclusions or grain boundaries cause locally increased stress levels resulting in the initiation of new microdefects. Each of these effects reduces the load-carrying capacity thus directly influencing the material's macroscopic mechanical response. In general, a crack's progression strongly depends on the state of stress which makes damage an essentially anisotropic[1] phenomenon. Furthermore, a material does not heal itself once the material's atomic bonds are broken off, except for living or biological materials for instance. That is why damage is also considered irreversible. This motivates the definition of *damage* being an, in general, anisotropic, irreversible process that describes the deterioration of a material's mechanical properties due to initiation, growth, and coalescence of microscopic defects (such as cracks or pores).

[1]Especially in the mechanics of solids, there are so-called *anisotropic* properties, whose response to external influences depends on the direction of those influences. Properties that do not show this directional dependency are referred to as *isotropic*.

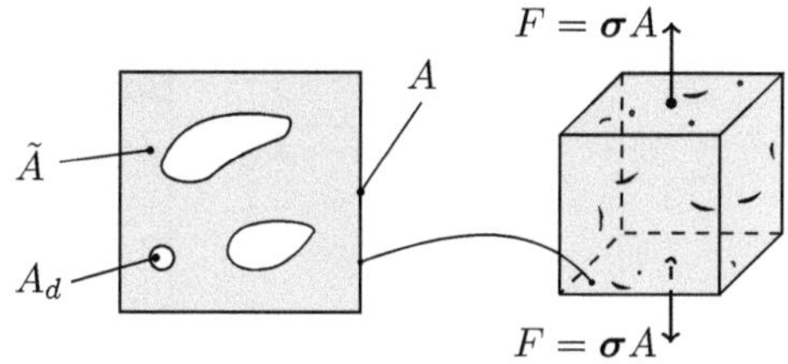

Fig. 1 Damaged RVE under uniaxial tension $F = \sigma A$. Due to the total area of micropores and -cracks A_d, the undamaged cross-sectional area A has been reduced to an effective one $\tilde{A}$ thus causing higher stresses

In its final stage, the material bonds are fully disintegrated and a macroscopic crack develops.[2]

1.2 *Basic Idea to Define an Isotropic Damage Variable*

This section outlines the basic ideas and main principles involved when introducing an isotropic damage variable. A general treatment of this subject from an engineering point of view can be found in [30, 35] e.g. Imagine a representative volume element (RVE) of a damaged material under uniaxial tension σ due to an acting force F as depicted in Fig. 1. Focussing on the cross-sectional view on the left of Fig. 1 an *effective cross section* $\tilde{A}$ is introduced by subtracting the damaged area A_d from the originally undamaged one A. The damage variable d is then defined as the ratio of damaged to undamaged area, i.e., $d := A_d/A$. The shape of the defining equation immediately suggests that d should only take values between zero and one, where zero refers to an undamaged and one to a completely damaged RVE, respectively. Owing to the reduced load-carrying area and the fact that stresses in this setting are essentially the forces divided by the area they act on, the same forces now cause higher so-called *effective stresses* $\tilde{\sigma} := F/\tilde{A}$ in the damaged material which can be related to the stresses σ via $\tilde{\sigma} = (1-d)^{-1}\sigma$ using only the defining equation for σ, $\tilde{\sigma}$, F, and d. To formulate the constitutive stress-to-strain relation, an additional hypothesis in form of the *principle of strain-equivalency* is introduced (see [30]). It states that any deformation behavior by a damaged material is described using the same constitutive laws as in the undamaged state but where the stresses are replaced with the effective stresses. As a consequence, the damaged state (as presented in Fig. 2) is mechanically equivalent to a ficticious undamaged state represented by $\tilde{\sigma}$, $\tilde{A}$, F see [35] e.g., and therefore deformations in both settings should be accompanied by the same strains. This gives rise to the introduction of an *effective Young's modulus* $\mathbb{E}_d = (1-d)\mathbb{E}$ and directly relates the damage variable to the change in Young's modulus via $d = 1 - \mathbb{E}_d\mathbb{E}^{-1}$.

[2]This description separates damage mechanics from fracture or crack mechanics, respectively, which focuses on crack propagation rather than initiation.

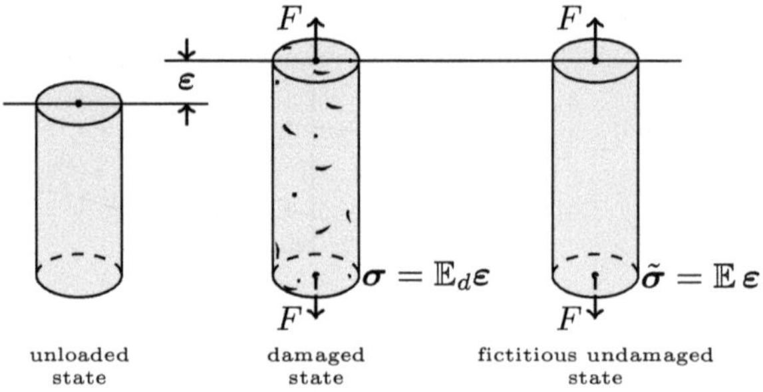

Fig. 2 Principle of strain equivalency. The strains ε in a damaged material are equal to those in a ficticious undamaged state

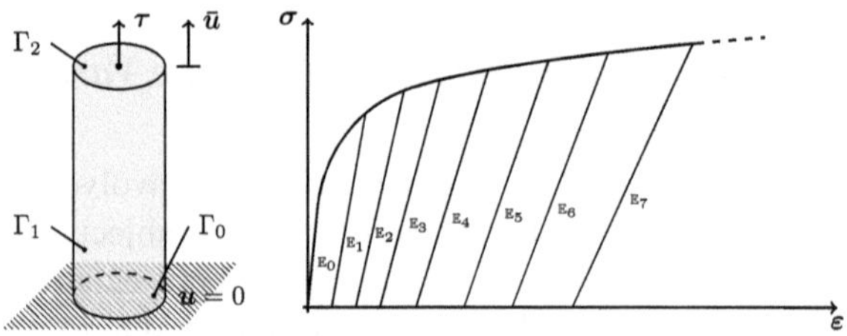

Fig. 3 Deterioration of Young's modulus $\mathbb{E}_0$ indicated through the slowly declining slope that represents the linear elastic behavior (right) of a specimen in a standard-tensile-like test (left)

1.3 Measuring Damage

Damage is *not directly accessible* to measurements. It is merely possible to measure the physical quantities it is connected to by modeling assumptions and, hence, interpretations strongly depend on that definition. In our case, where the concept of effective stress and strain-equivalency are employed, the measurements are essentially connected to the coupling between deformation and damage as pointed out in [30, 31]. Most commonly, the change in a material's elastic behavior expressed via the deterioration of Young's modulus is used to retrieve information on the damage state. Usually, this parameter is computed from data obtained by a setup similar to the tensile test in material science. Fig. 3 shows the specimen of such a test idealized as a cylinder, where the parts of its surface are named according to their respective parts in the model presented in Sect. 2. The specimen is clamped at its bottom end and while the lateral surface Γ_1 is tension free, a force in the shape of a tension $\boldsymbol{\tau}$ is applied on its top end. Its elongation is then measured via the boundary displacements $\bar{\boldsymbol{u}}$. The proposed experiment is then conducted by slowly loading and unloading the specimen, where the load gradually gains in magnitude. This causes a deterioration in Young's modulus indicated by the lessened slope in the elastic behavior due to evolving damage. This change is depicted in the stress-over-strain diagram in Fig. 3 as well. The earlier presented relation between damage and the change in the elasticity constant (see Sect. 1.2) provides the means to predict the current damage state.

1.4 Survey on Damage Models

The notion of damage was introduced first by Kachanov in 1958 and was later extended by Rabotnov (see [35, Sect. 2.1.1.] and the references therein).[3]
From an engineering perspective, an extensive survey on advances in the field of damage mechanics can be found in [41], and monographs are [30, 35], e.g.

The first model that captured a broad attention among mathematicians was proposed in [16, 17] by Frémond and Nedjar in 1996 and describes isotropic damage in a quasi-static linear elastic setting. Its popularity is mainly caused by the incorporation of the *gradient of damage* that has, from a mathematical viewpoint, a compactifying character and opens up opportunities for a successful analysis (see [33], e.g.). The basic idea of deriving Frémond's model lies in a generalization of the *principle of virtual power* (see [30, 36, Chap. 2], e.g.) by introducing the power of microscopic forces to account for microscopic movements. The constitutive equations are derived from specifying the free energy and a pseudo-potential of dissipation incorporating the gradient of damage to take into account the damaging effect, which an already damaged region has on its neighborhood. A momentum balance equation coupled with a damage evolution equation together with suitable boundary and initial conditions are deduced. In particular, a homogeneous Neumann type boundary condition for the damage variable is assumed. The model is complemented by two additional constraints enforcing the damage value to remain between zero and one on the one hand and for its rate to remain positive due to the irreversibility of damage on the other. These constraints are enforced using subdifferentials and lead to a system of quasi-elliptic partial differential equations for the momentum balance and a parabolic differential inclusion for the damage evolution. The right-hand side of that inclusion is a constitutive function describing the *damage source* resulting from tension. Later contributions include more general source terms.

The first analytic result for Frémond's model was established in [15], where the authors consider a one-dimensional visco-elastic material subjected to isotropic partial[4] damage. Local-in-time existence of a weak solution was proven with the help of an auxiliary problem attained through various transformations, truncation and regularization procedures which then tends to a solution of the original problem in the limit. To this end, they strongly rely on embedding theorems. Uniqueness is established under the simplifying assumption of damage being a reversible process.

---

[3] Analytic results of the Kachanov-Rabotnov model are discussed in [1, 38], e.g.

[4] Partial damage refers to a damage state $0 \le d \le \delta_0 < 1$, where δ_0 denotes the maximal damage possible. Partial damage models are more feasible for mathematical analysis, because they preserve uniform coercivity of the effective elasticity tensor, whereas complete damage leads to a degenerate momentum balance equation.

The authors of [5] proved a local-in-time existence and uniqueness result for a three-dimensional damage-rate-regularized version of Frémond's model in a quasi-static setting with partial damage in 2004. They use a truncation argument and the Yosida approximation to introduce an auxiliary problem, whose weak solvability follows from a fixed-point argument. A limit analysis then shows the existence of a solution to the original truncated system. In the end, the truncation can be dropped for a small-enough time interval. The proof strongly relies on the aforementioned regularization. In [4], a subsequent article, Bonetti et al. analyze a similar scenario but this time in a dynamic setting with a visco-elastic material and the original damage evolution law from [16]. The role of the regularizing effect stems from the viscous term in the momentum balance equation. A similar reasoning as in [5] is applied, but stronger assumptions for a unique solution are necessary. The analysis of a more complex model considering temperature dependency and accounting for microscopic accelerations, too, is presented in [3].

In 2005, [28] studied the quasi-static evolution of *reversible* isotropic partial damage in an elastic-visco-plastic material. Kuttler was able to prove existence and uniqueness of a local-in-time weak solution. He circumvented the necessity of a subgradient in the damage evolution that enforces the smallness assumption on the damage variable by demanding additional structure for the damage source function. This lead to stronger regularity results for the solution.[5] In many parts based on this work, Kuttler and Shillor analyze a quasi-static evolution of reversible isotropic partial damage in an elastic body in [27]. While the aforementioned additional assumptions on the damage source were dropped, the authors were able to show existence and uniqueness of a global-in-time weak solution under moderate restrictions on the data by employing a fixed-point argument.[6]

Another important work and starting point for further research on damage models is [34], where the author analyzes the evolution of rate-independent systems in general. These are systems, whose overall response does not change due to a strictly monotone time reparametrization, which, in the context of damage evolution, translates to a positive homogeneous dissipation potential for damage of degree one. From an analytic perspective, the main contribution in [34] is the introduction of another in some way weaker notion of a solution, so-called *energetic solutions*, on which we want to comment briefly in the following. In general, evolution processes can be written in the form of evolutionary quasi-variational inequalities or (equivalently) as differential inclusions using subdifferentials. However, the basic idea for the *energetic formulation* is to look for a process to satisfy the overall stored energy and the dissipation (pseudo-) potential previously used to derive the aforementioned variational inequality. One may consider this formulation weaker since it is derivative free for the evolving variable as well as for the participating functionals. This means, that

[5]This approach was also continued in some subsequent publications of the author, e.g. in the analysis of a contact problem in an otherwise equal setting in [8] or a dynamic frictional contact problem of a visco-elastic material with damage in [7]. Both these articles further developed the ideas presented in [10].

[6]For numeric investigations and results see [9].

this form does not require smooth solutions in time or space and thus being more adequate for many mechanical systems (see [32]). However, the regularity assumptions on the loadings has to be a little higher, because the power of the external forces is given via the partial time derivative of the energy storage potential and has to be well-defined, see [34]. In the context of rate-independent damage modeling, this approach was firstly used in [33] to investigate partial and complete isotropic damage in nonlinear elasticity. In view of partial isotropic damage, [40] generalized these results and proved better spatial regularity properties for the damage variable (see also [39]). Under certain convexity assumptions on the energy functional, the authors were also able to improve temporal regularity. In [2] the authors focus on solvability of a complete damage problem in linear elasticity. Also temporal regularity properties of the solutions were shown. For rate-independent damage coupled with rate-dependent effects for the displacements see [29, 37] and the references therein. Fully rate-independent systems while neglecting the gradient of damage were investigated in [13, 14].

In [22], Heinemann and Kraus extended the framework of Cahn Hillard systems, a typical phase field approach for describing phase separation and coarsening phenomena in alloys, by incorporating rate-dependent damage processes. A suitable weak formulation involving variational inequalities and a total energy-dissipation inequality as well as an energy estimate inspired by the concept of energetic solutions was introduced. This approach proved very fruitful. It was the basis for many subsequent publications concerning phase separation coupled with damage, see [23, 26] for a quasi-static setting and [21] for a dynamic one. Also in the just linear elastic case (quasi-static or dynamic) without considering phase separation the outlined notions were used to investigate solvability in [19, 20].

Optimal control problems incorporating damage are analyzed in [12, 25]. Both articles take the surface traction acting on a part of the boundary of an elastic body as a boundary control. In this first publication, the authors consider a quasi-static linear elastic setting with reversible damage. They show solvability of this problem in a class of weak variational solutions using the convergence concept in variable spaces. In [12], the authors consider a two-dimensional visco-elastic material and model damage via a phase field approach. The cost functional penalizes maximal deviations from prescribed damage profiles.

To the best of our knowledge, there are no analytic works on parameter identification in elastic damaged continua. In view of this, well-posedness of the presented models still remains open questions. In some cases, uniqueness and stability were only proven for reversible damage. The fact that many solutions are just local in time poses another limitation, because we are generally interested in solutions for finite but adjustable long time spans.

At the cost of neglecting the gradient of damage in the model we propose in the following Sect. 2, we were able to show that the direct problem is well-posed. We feel that this is an acceptable price to pay, because there is no real consensus in the engineering community on whether its inclusion is generally leading to physically more accurate models. Both scenarios are topic of ongoing research in the engineering community.

2 The Model Based on Kachanov's Approach

Under the assumption of partial isotropic damage in a dynamic linear elastic setting, we propose the following damage model.

$$\sigma(t,\boldsymbol{x}) = (1-d(t,\boldsymbol{x}))\mathbb{E}(\boldsymbol{x})\varepsilon(\boldsymbol{u}(t,\boldsymbol{x})) \quad \text{in } S\times\Omega, \tag{1a}$$
$$\boldsymbol{u}''(t,\boldsymbol{x}) - \operatorname{div}(\sigma)(t,\boldsymbol{x}) = \boldsymbol{f}(t,\boldsymbol{x}) \quad \text{in } S\times\Omega, \tag{1b}$$
$$d'(t,\boldsymbol{x}) = (1-d(t,\boldsymbol{x}))^{-\alpha} g(t,\boldsymbol{x},\nabla\boldsymbol{u}(t,\boldsymbol{x})) \quad \text{in } S\times\Omega, \tag{1c}$$
$$\boldsymbol{u}(t,\boldsymbol{x}) = \boldsymbol{0} \quad \text{on } S\times\Gamma_0, \tag{1d}$$
$$\sigma(t,\boldsymbol{x})\boldsymbol{\nu}(t,\boldsymbol{x}) = \boldsymbol{0} \quad \text{on } S\times\Gamma_1, \tag{1e}$$

with either Dirichlet

$$\boldsymbol{u}(t,\boldsymbol{x}) = \bar{\boldsymbol{u}}(t,\boldsymbol{x}) \quad \text{on } S\times\Gamma_2 \tag{1f}$$

or Neumann type boundary conditions

$$\sigma(t,\boldsymbol{x})\boldsymbol{\nu}(t,\boldsymbol{x}) = \boldsymbol{\tau}(t,\boldsymbol{x}) \quad \text{on } S\times\Gamma_2, \tag{1g}$$

on Γ_2 and initial values

$$\boldsymbol{u}(0,\boldsymbol{x}) = \boldsymbol{u}_0(\boldsymbol{x}), \quad \boldsymbol{u}'(0,\boldsymbol{x}) = \boldsymbol{u}_1(\boldsymbol{x}), \quad d(0,\boldsymbol{x}) = d_0(\boldsymbol{x}) \quad \text{in } \Omega. \tag{1h}$$

Additionally, the damage variable should also comply with

$$0 \le d(t,\boldsymbol{x}) \le \delta_1 < 1, \quad d'(t,\boldsymbol{x}) \ge 0 \quad \text{in } S\times\Omega, \tag{1i}$$

for some constant δ_1. The domain $\Omega \subset \mathbb{R}^n$ represents the solid under consideration during a time interval $S := (0,T)$ where $0 < T < \infty$. The *momentum balance equation* in (1b) describes a material's dynamic mechanical response to a specific volumetric loading $\boldsymbol{f}$ (see [11], e.g.), where the displacements are denoted by $\boldsymbol{u}$ and the second Piola-Kirchhoff stresses by σ. By only considering small deformations and employing the aforementioned principles, we retrieve the linear elastic law for a damaged continua in (1a) where ε represents the linearized Cauchy-Green strains and $\mathbb{E}$ the non-homogeneous elasticity tensor. This equation couples the mechanical part of our model to the damage evolution equation (1c), which is, together with the constraints in (1i), based on Kachanov's approach. Here, d denotes the damage variable and $\alpha \ge 1$ some fixed material parameter. The function g is the *damage supply* modeling the effective stresses, whose shape determines, how the relation between deformation and damage comes into play.[7] The constraints in (1i) are owed to the irreversibility and boundedness of damage where the constant δ_1 ensures a partial damage setting. The boundary conditions in (1d) – (1g) are taken from the

[7]This is why, as a long-term project, we aim to identify g. The only assumptions on g stem from the regularity needed to perform the analysis (see [18]).

experimental setup, where Γ_0 denotes the clamped bottom end of the specimen, Γ_1 its tension free lateral surface, and Γ_2 its top end, where either the displacements $\bar{\boldsymbol{u}}$ or the forces $\boldsymbol{\tau}$ are prescribed (cf. to Fig. 3).

Under reasonable assumptions on the data and the structure of g, we were able to prove global-in-time existence and uniqueness of a solution to the problems presented in (1) (see [18] for details), which Lipschitz continuously depend on the provided data. These results allow the introduction of parameter-to-state operators $\Lambda_{\bar{\boldsymbol{u}}}$, $\Lambda_{\boldsymbol{\tau}}$ mapping the parameter g to either the measured displacements $\bar{\boldsymbol{u}}$ or the traction $\boldsymbol{\tau}$, respectively. The well-posedness of this *direct problem* is an immediate consequence of these results. The next step is a rigorous analysis of the inverse problem.

Acknowledgements The authors are indebted to Michael Böhm for initiating and supporting this research.

References

1. Altenbach, H., Deuring, P., Naumenko, K.: A system of ordinary and partial differential equations describing creep behavior of thin-walled shells. Z. Ana. Anw. **18**, 1003–1030 (1999)
2. Bouchitté, G., Mielke, A., Roubíček, T.: A complete-damage problem at small strains. Z. Angew. Math. Phys **60**, 205–236 (2009)
3. Bonetti, E., Bonfanti, G.: Well-posedness results for a model of damage in thermoviscoelastic materials. Ann. Inst. Henri Poincaré, Anal. Non Linéaire **25**, 1187–1208 (2008)
4. Bonetti, E., Schimperna, G., Segatti, A.: On a doubly nonlinear model for the evolution of damaging in viscoelastic materials. J. Differ. Equ. **218**, 91–116 (2005)
5. Bonetti, E., Schimperna, G.: Local existence for Frémond's model of damage in elastic materials. Contin. Mech. Thermodyn. **16**, 319–335 (2004)
6. Bonetti, E., Frémond, M.: Damage theory: microscopic effects of vanishing macroscopic motions. Comput. Appl. Math. **22**, 313–333 (2003)
7. Campo, M.A., Fernández, J.R., Kuttler, K.L.: Analysis of a dynamical frictional contact problem with damage. Fin. El. Anal. Des. **45**, 659–674 (2009)
8. Campo, M.A., Fernández, J.R., Kuttler, K.L.: An elastic-viscoplastic quasistatic contact problem with damage. Comput. Methods Appl. Mech. Engrg **196**, 3219–2339 (2007)
9. Campo, M.A., Fernández, J.R., Kuttler, K.L., Shillor, M.: Quasistatic evolution of damage in an elastic body: numerical analysis and computational experiments. Appl. Numer. Math. **57**, 976–988 (2007)
10. Campo, M.A., Fernández, J.R., Han, W., Sofonea, M.: A dynamic viscoelastic contact problem with normal compliance and damage. Fin. Elem. Anal. Des. **42**, 1–24 (2005)
11. Ciarlet, P.: Mathematical Elasticity, vol. I. North-Holland, Amsterdam (1988)
12. Farshbaf-Shaker, M.H., Heinemann, C.: A phase field approach for optimal boundary control of damage processes in two-dimensional viscoelastic media. Math. Models Methods Appl. Sci. **25**, 2749–2793 (2015)
13. Fiaschi, A., Knees, D., Stefanelli, U.: Young-measure quasi-static damage evolution. Arch. Rational Mech. Anal. **203**, 415–453 (2012)
14. Francfort, G.A., Garroni, A.: A variational view of partial brittle damage evolution. Arch. Ration. Mech. Anal. **182**, 125–152 (2006)
15. Frémond, M., Kuttler, K.L., Shillor, M.: Existence and uniqueness of solutions for a dynamic one-dimensional damage model. J. Math. Anal. Appl. **229**, 271–294 (1999)
16. Frémond, M., Nedjar, B.: Damage, gradient of damage and principle of virtual power. Int. J. Sol. Struct. **33**, 1083–1103 (1996)

17. Frémond, M., Nedjar, B.: Damage in concrete: the unilateral phenomenon. Nucl. Eng. Des. **156**, 323–335 (1996)
18. Grützner, S.: An approach to parameter identification in damaged continua. Diploma thesis, University of Bremen (2015)
19. Heinemann, C., Kraus, C.: Complete damage in linear elastic materials: modeling, weak formulation and existence results. Calc. Var. Partial Differ. Equ. **54**, 217–250 (2015)
20. Heinemann, C., Kraus, C.: Existence of weak solutions for a hyperbolic-parabolic phase field system with mixed boundary conditions on nonsmooth domains. SIAM J. Math. Anal. **47**, 2044–2073 (2015)
21. Heinemann, C., Kraus, C.: Existence of weak solutions for a pde system describing phase separation and damage processes including inertial effects. Discrete Contin. Dynam. Syst. **35**, 2565–2590 (2015)
22. Heinemann, C., Kraus, C.: Degenerating Cahn-Hilliard systems coupled with mechanical effects and complete damage processes. Math. Bohem. **139**, 315–331 (2014)
23. Heinemann, C., Kraus, C.: Existence results for diffuse interface models describing phase separation and damage. Eur. J. Appl. Math **24**, 179–211 (2013)
24. Heinemann, C., Kraus, C.: Existence of weak solutions for Cahn-Hilliard systems coupled with elasticity and damage. Adv. Math. Sci. Appl. **21**, 321–359 (2011)
25. Kogut, P.I., Leugering, G.: Optimal and approximate boundary controls of an elastic body with quasistatic evolution of damage. Math. Methods Appl. Sci. **38**, 2739–2760 (2015)
26. Kraus, C., Bonetti, E., Heinemann, C., Segatti, A.: Modeling and analysis of a phase field system for damage in phase separation processes in solids. J. Differ. Equ. **258**, 3928–3959 (2015)
27. Kuttler, K.L., Shillor, M.: Quasistatic evolution of damage in an elastic body. Nonlinear Anal. Real World Appl. **7**, 674–699 (2006)
28. Kuttler, K.L.: Quasistatic evolution of damage in an elastic-viscoplastic material. Elec. J. Differ. Equ. **2005**, 1–25 (2005)
29. Lazzaroni, G., Rossi, R., Thomas, M., Toader, R.: Some remarks on a model for rate-independent damage in thermo-visco-elastodynamics. J. Phys. Conf. Ser. **727** (2016)
30. Lemaître, J., Chaboche, J.-.L: Mechanics of Solid Materials. Cambridge University Press, New York (1990)
31. Lemaître, J., Dufailly, J.: Damage measurements. Eng. Fract. Mech. **28**, 643–661 (1987)
32. Mainik, A., Mielke, A.: Existence results for energetic models for rate-independent systems. Calc. Var. Partial Differ. Equ. **22**, 73–99 (2005)
33. Mielke, A., Roubíček, T.: Rate-independent damage processes in nonlinear elasticity. Math. Mod. Methods Appl. Sci. **16**, 177–209 (2006)
34. Mielke, A.: Evolution of rate-independent systems. In: Dafermos, C.M., Feireisl, E. (eds.) Handbook of Differential Equations. 461–559. Elsevier, Amsterdam (2005)
35. Murakami, S.: Continuum Damage Mechanics. Springer, Heidelberg (2012)
36. Nedjar, B.: Damage and gradient of damage in transient dynamics. In: IUTAM Argoul, P., Frémond, M., Nguyen, Q.S. (eds.) IUTAM Symposium on Variations of Domain and Free-Boundary Problems in Solid Mechanics. Proceedings of the IUTAM Symosium held in Paris, France, April 22-25, 1997; 189–196. Springer, Berlin (1999)
37. Roubíček, T., Tomassetti, G.: Thermomechanics of damageable materials under diffusion: modelling and analysis. Z. Angew. Math. Phys. **66**, 3535–3572 (2016)
38. Shutov, A.V., Sändig, A.-M.: Mathematical analysis of fully coupled approach to creep damage. arXiv:math-ph/0601052v1 (2006)
39. Thomas, M.: Quasistatic damage evolution with spatial BV-regularization. Disc. Cont. Dyn. Syst. **6**, 235–255 (2013)
40. Thomas, M., Mielke, A.: Damage of nonlinearly elastic materials at small strain - existence and regularity results. Z. Angew. Math. Mech. **90**, 88–112 (2010)
41. Voyiadjis, G.Z., Kattan, P.I.: Advances in Damage Mechanics. Elsevier, Amsterdam (2006)

Structured Deformations of Continua: Theory and Applications

Marco Morandotti

Abstract The scope of this contribution is to present an overview of the theory of structured deformations of continua, together with some applications. Structured deformations aim at being a unified theory in which elastic and plastic behaviours, as well as fractures and defects can be described in a single setting. Since its introduction in the scientific community of rational mechanicists [10], the theory has been put in the framework of variational calculus [8], thus allowing for solution of problems via energy minimization. Some background, three problems and a discussion on future directions are presented.

1 Introduction

Many and complex are the deformations of a body when it responds to external loading. In the last century, and even before, several theories have been proposed to study phenomena such as elasticity, plasticity, fracture...More recently, both the advent of modern instruments capable of resolving the core of the matter, and more powerful computers allowed scientists to bridge different length scales together and to formulate theories that range from the subatomic level to the macroscopic one. In this way, phenomena that were described from a macroscopic viewpoint started being related to their microscopic counterpart: it was possible to model at the atomistic level and to explain phenomena that are experienced at the macroscopic scale. An example of this is the understanding that the motion and pile-up of defects such as dislocations is the microscopic process which is responsible for plasticity.

The necessity of having a theoretical tool that permits inclusion of multiple scales is evident. *Structured deformations* [10] (see also [3]) respond to this need by providing a multiscale geometry that captures the contributions at the macroscopic level of both smooth geometrical changes and non-smooth geometrical changes at submacroscopic levels. These non-smooth geometrical changes, which are called

M. Morandotti (✉)
Technische Universität München, 85748 Garching, Germany
e-mail: marco.morandotti@ma.tum.de

P. van Meurs et al. (eds.), *Mathematical Analysis of Continuum Mechanics and Industrial Applications II*, Mathematics for Industry 30,
https://doi.org/10.1007/978-981-10-6283-4_11

disarrangements, encode the presence of cracks and defects in the continuum. The geometrical effects at the macrolevel of submicroscopic disarrangements are captured by sequences of approximating, piecewise smooth deformations f_n that converge to the macroscopic deformation field g and whose gradients ∇f_n converge to a field G that might differ from ∇g. The jumps in f_n contribute to the interfacial part of an initial energy response $E(f_n)$, they diffuse throughout portions of the body, and they contribute in the limit to both the bulk and interfacial parts of a relaxed energy response $I(g, G)$. This is the main result contained in the approximation and relaxation theorems [10, Theorem 5.8] and [8, Theorems 2.12 and 2.17], and is the novelty of the theory. The main feature of the relaxation theorems is that they provide a representation formula for the relaxed energy functional which can then be expressed as an integral. The relaxed bulk and surface energy densities are obtained through the so-called cell formulas, which are expressed as minimum problems.

Structured deformations have been successfully applied in many contexts to model the deformation of bodies that also include plastic deformations and cracks [7, 11–13]. The theory has been extended to more general contexts, especially by defining second-order structured deformations [19], which permit the inclusion of bending effects in the energy functional. Applications and developments of second-order structured deformations are provided, for example, in [5, 18, 21]. Also relevant are the works [4, 20, 23], which focus on interfacial energies, relevant, among other things, for the study of granular and composite materials (see [16] in this context), as well as [22], where a more general functional setting is investigated.

In the present note, we shall present the basic definitions of structured deformations and the relaxation theorems contained in [8, 10], and the applications contained in [4, 6, 16], which contain contributions of the author.

In Sect. 2, we present the general mechanical and functional setting, as well as define the energies that are interesting to our problems. In Sect. 3, we present the results for three specific problems. Finally, in Sect. 4, we outline some future directions.

2 General Setting and Energies

In this section we recall the definition of structured deformation and also present the main results about the relaxation of non-convex energies.

Always in this note, and unless differently specified, Ω will denote a bounded open subset of the N-dimensional Euclidean space $\mathbb{R}^N$.

2.1 Functional Setting

We start by defining the function spaces which are relevant in our context. We assume that the reader is familiar with the Lebesgue spaces $L^p(\Omega; \mathbb{R}^d)$ ($p \in [1, \infty]$), with the space of Radon measures $\mathcal{M}(\Omega;\mathbb{R}^d)$, and with the different notions of convergence (strong, weak, and weak-*).

Definition 1 (*see* [2]) The space of $\mathbb{R}^d$-valued functions of *bounded variation* is defined as $BV(\Omega;\mathbb{R}^d) := \{u \in L^1(\Omega;\mathbb{R}^d) : Du \in \mathcal{M}(\Omega;\mathbb{R}^{d\times N})\}$. The distributional derivative Du of a BV function is characterized by $Du = \nabla u\mathcal{L}^N + [u]\otimes \nu\mathcal{H}^{N-1}\llcorner S(u) + D^c u$, where ∇u is the approximate gradient (see [2, Definiton 3.70 and Theorem 3.83]), $[u] := u^+ - u^-$ is the jump of u across the jump set $S(u)$ (which can be proved to be $(N-1)$-rectifiable), ν is the normal to $S(u)$, and $D^c u$ is the *Cantor part* of the measure Du. The space of *special functions of bounded variation* is defined as

$$SBV(\Omega;\mathbb{R}^d) := \{u \in BV(\Omega;\mathbb{R}^d) : D^c u = 0\}, \tag{1}$$

the set of BV functions for which the singular part of Du is reduced to the jump part.

We now present the two definitions of structured deformation of [10] and [8].

Definition 2 (*see* [10]) A *structured deformation* is a triple (κ, g, G), where κ is a surface-like subset of Ω, and the injective and piecewise differentiable map $g\colon \Omega \to \mathbb{R}^N$ the piecewise continuous tensor field $G : \Omega \to \mathbb{R}^{N\times N}$ are such that

$$0 < C < \det G(x) \leq \det \nabla g(x) \qquad \text{at each point } x \in \Omega.$$

In view of Definition 2, κ describes preexisting, unopened macroscopic cracks and the map g and its classical gradient ∇g describe macroscopic changes in the geometry of the body. A geometrical interpretation of the field G is provided by the following approximation theorem.

Theorem 1 (see [10, Theorem 5.8]) *For each structured deformation* (κ, g, G) *there exists a sequence of injective, piecewise smooth deformations* f_n *and a sequence of surface-like subsets* κ_n *of the body such that*

$$g = \lim_{n\to\infty} f_n, \qquad G = \lim_{n\to\infty} \nabla f_n, \qquad \textit{and} \qquad \kappa = \bigcup_{n=1}^{\infty}\bigcap_{j=n}^{\infty} \kappa_j. \tag{2}$$

The limits in (2) are taken in the sense of L^∞convergence. From (2) we see that G captures the effects at the macroscopic level of smooth geometrical changes at submacroscopic levels; G is usually referred to as the *deformation without disarrangements*.

The space $SBV(\Omega;\mathbb{R}^d)$ defined in (1) formalizes the notion of discontinuity of a function u with the introduction of the jump set $S(u)$. In this way, one can think of linking the role of κ with that of g in Definition 2, by considering pairs made of a deformation g, also carrying the information about the cracks, and a matrix-valued field G.

Definition 3 (*see* [8]) The space of structured deformations is defined as

$$SD(\Omega) := \{(g, G) : g \in SBV(\Omega; \mathbb{R}^d), G \in L^1(\Omega; \mathbb{R}^{d\times N})\}. \tag{3}$$

In view of Definition 3, Theorem 1 has the following counterpart.

Theorem 2 (see [8, Theorem 2.12]) *Let* $(g, G) \in SD(\Omega)$. *Then there exist* $u_n \in SBV(\Omega; \mathbb{R}^d)$ *such that*

$$u_n \to g \text{ in } L^1(\Omega; \mathbb{R}^d), \qquad \nabla u_n \overset{*}{\rightharpoonup} G \text{ in } \mathcal{M}(\Omega; \mathbb{R}^{d\times N}). \tag{4}$$

The proof of Theorem 2 can be achieved by combining Alberti's theorem [1, Theorem 3] and an approximation result in BV by piecewise constant functions (see, *e.g.*, [8, Lemma 2.9]). It is worth mentioning that an alternative proof of a weakerversion of the approximation theorem, without using Alberti's theorem, is proposed in [22, Theorem 7.1]. We want to stress that the convergences in (4) allow for the limit of a sequence of functions and the limit of its gradients to be unrelated, and in fact there is no *a priori* relationship between g and G. Even more importantly, the *disarrangements tensor* defined by the difference

$$M := \nabla g - G \tag{5}$$

has a fundamental geometrical meaning: it captures, in the limit as $n \to \infty$, the volume density of separations and slips between pieces of the body, and how this is determined by the interfacial discontinuities of the approximating deformations u_n in Theorem 2 (or f_n in Theorem 1). The tensor M can also be interpreted as a measure of how non classical a deformation is: $M = 0$ corresponds to $\nabla g = G$, and therefore the convergences in (2) and (4) hold in a stronger sense. In this regard, from (5) and (2) one trivially obtains $M = \nabla\big(\lim_{n\to\infty f_n}\big) - \lim_{n\to\infty} \nabla f_n$, which can be considered a quantitative measure of the lack of commutativity of the classical gradient and the limit operator in the L^∞ convergence.

The tensor M and its derivatives, such as curl M, are fundamental quantities to describe the presence of defects such as dislocations, and how they intervene in the response of solids (see the discussion in [4, Sect. 1.1]).

We close this discussion by presenting two examples (see [8, 10])

Example 1 (*The broken ramp*) Let $N = 1$, $\Omega = (0, 1)$, $\kappa = \emptyset$, $g(x) = 2x$, and $G(x) = 1$. An approximating sequence is given by $f_n(x) := x + \frac{k}{n}$, for $\frac{k}{n} \leq x < \frac{k+1}{n}$ and $k = 0, \dots, n-1$. Indeed, $f_n(x) \to 2x$ as $n \to \infty$, and $\nabla f_n(x) = 1$ for every n, so that (2) and (4) are satisfied; yet, the distributional derivative is given by $Df_n = 1 + \sum_{k=1}^{n-1} \frac{1}{k}\delta_{k/n}$ and it shows the emergence of jumps discontinuities, which become smaller and smaller in magnitude and diffuse in the whole domain as $n \to \infty$, see Fig. 1.

Example 2 (*The deck of cards*) In this example in $N = 3$ dimensions, a simple shear of a cubic domain $\Omega = (0, 1)^3$ is considered. Let $\kappa = \emptyset$, $g(x) = (x_1 + x_3, x_2, x_3)$,

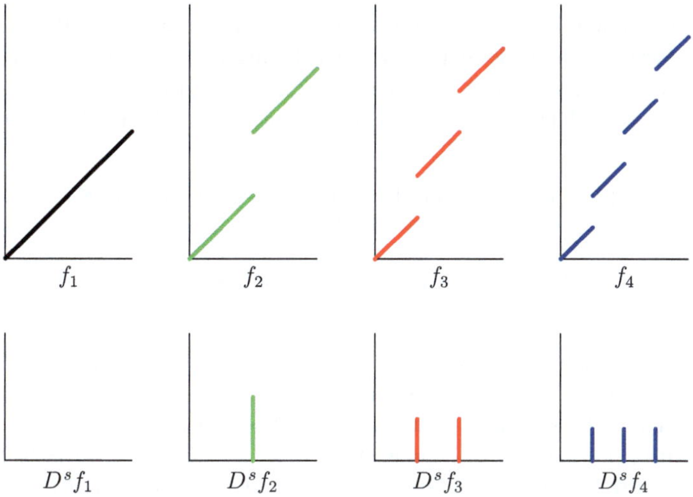

Fig. 1 The first four steps of the approximation in Example 1. The top row depicts the approximants $f_1, \dots, f_4$; the bottom row depicts the singular parts $D^s f_1, \dots, D^s f_4$

and $G(x) = \mathbb{I}$. An approximating sequence is given by $f_n(x) := \big(x_1 + \frac{k}{n}, x_2, x_3\big)$, for $\frac{k}{n} \leq x_3 < \frac{k+1}{n}$ and $k = 0, \dots, n-1$. Then, $f_n(x) \to g(x)$ as $n \to \infty$, and $\nabla f_n(x) = \mathbb{I}$ for every n, so that (2) and (4) are once again satisfied; here, $Df_n = \mathbb{I} + \sum_{k=1}^{n-1} \frac{1}{k} \delta_{k/n}(x_3) e_1 \otimes e_3$, and also in this case the jumps diffuse to the bulk.

The effect that the singular part of Du_n diffuses to the bulk is crucial to understanding structured deformations and the cell formulas in the relaxation theorems that we will present later. What is happening is that, as $n \to \infty$, the singular sets diffuse ($\mathscr{H}^{N-1}(S(u_n)) \to +\infty$) in a controlled way, which is measured by the boundedness of the total variation of the singular part of the measure, $|D^s u_n|(\Omega) < +\infty$.

2.2 *Energies and Relaxation*

We present here some general energy functionals for deformations of solids, as well as the relaxation theorems to define functionals on the space $SD(\Omega)$ introduced in (3). Energy minimization is a convenient way to find equilibrium configurations of a body undergoing internal stresses and external loadings. In the general case, the energy densities are not convex, or quasi-convex, so that the energy landscape can show many local minima. In calculus of variations, this situation is resolved by computing the (quasi-)convex envelope of the energy densities, therefore obtaining the relaxed functional, that is the largest lower semicontinuous functional below the given one. We refer the reader to treatises on calculus of variations for the relationship between lower semicontinuity and convexity (see, *e.g.*, [14]).

Let now $u \in SBV(\Omega;\mathbb{R}^d)$ and consider an energy functional defined by

$$E(u) := \int_\Omega W(\nabla u(x))\,\mathrm{d}x + \int_{S(u)\cap\Omega} \Psi([u],\nu(u))\,\mathrm{d}\mathscr{H}^{N-1}, \tag{6}$$

where the desities $W\colon \mathbb{R}^{d\times N} \to [0,+\infty)$ and $\Psi\colon \mathbb{R}^d \times \mathbb{S}N-1 \to [0,+\infty)$ satisfy:

(H_1) $\exists\, C>0$ such that $\forall\, A,B \in \mathbb{R}^{d\times N}$: $|W(A)-W(B)| \le C|A-B|(1+|A|^{p-1}+|B|^{p-1})$ for some $p\in(1,+\infty)$;
(H_2) $\exists\, c_1, C_1 > 0$ such that $\forall\,(\lambda,\nu) \in \mathbb{R}^d \times \mathbb{S}N-1$: $c_1|\lambda| \le \Psi(\lambda,\nu) \le C_1|\lambda|$;
(H_3) (positive homogenity of degree 1) $\forall\,\lambda\in\mathbb{R}^d$, $t>0$: $\Psi(t\lambda,\nu) = t\Psi(\lambda,\nu)$;
(H_4) (subadditivity) $\forall\,\lambda_1,\lambda_2 \in \mathbb{R}^d$: $\Psi(\lambda_1+\lambda_2,\nu) \le \Psi(\lambda_1,\nu)+\Psi(\lambda_2,\nu)$.

Given a structured deformation $(g,G) \in SD(\Omega)$, the relaxation of the energy (6) is

$$\begin{aligned} I(g,G) := \inf\Big\{ &\liminf_{n\to\infty} E(u_n) : u_n \in SBV(\Omega;\mathbb{R}^d),\ u_n \to g \text{ in } L^1(\Omega;\mathbb{R}^d), \\ &\nabla u_n \overset{*}{\rightharpoonup} G \text{ in } \mathscr{M}(\Omega;\mathbb{R}^{d\times N}),\ \sup_n \|\nabla u_n\|_{L^p} < \infty \Big\}. \end{aligned} \tag{7}$$

Remark 1 We notice that the lack of coercivity of the bulk density W is compensated by the boundedness constraint on the approximating functions u_n requested in (7). This gives the freedom to treat more general densities W satisfying (H_1). A strategy to circumvent the lack of coercivity is proposed in [8, proof of Proposition 2.22, Step 2]: one first adds a fictitious coercivity $W^\varepsilon(\cdot) := W(\cdot) + \varepsilon|\cdot|^p$, and then lets $\varepsilon \to 0$. The requirement in (H_2) of coercivity for Ψ in the first variable can be removed [8, Remark 3.3] by modifying the requirements in (H_2) and (H_3) and the boundedness requirement in (7).

Let us denote by $Q := (-\frac12,\frac12)^N$ the unit cube in $\mathbb{R}^N$ and by Q_ν the rotated one such that two faces are perpendicular to the vector $\nu \in \mathbb{S}N-1$. The main result regarding the relaxed energy in (7) is the following.

Theorem 3 (see [8, Theorem 2.17]) *Let $(g,G) \in SD(\Omega)$ with $G \in L^p(\Omega;\mathbb{R}^{d\times N})$, and let the bulk and surface energy densities W and Ψ satisfy (H_1)-(H_4). Then*

$$I(g,G) = \int_\Omega H(\nabla g, G)\,\mathrm{d}x + \int_{S(g)\cap\Omega} h([g],\nu(g))\,\mathrm{d}\mathscr{H}^{N-1},$$

where, for $A,B \in \mathbb{R}^{d\times N}$, $\lambda \in \mathbb{R}^d$, $\nu \in \mathbb{S}N-1$

$$H(A,B) := \inf\left\{ \int_Q W(\nabla u)\,\mathrm{d}x + \int_{S(u)\cap Q} \Psi([u],\nu(u))\,\mathrm{d}\mathcal{H}^{N-1} : \right.$$
$$\left. u \in SBV(Q,\mathbb{R}^d),\ u|_{\partial Q} = Ax,\ |\nabla u| \in L^p(\Omega),\ \int_Q \nabla u\,\mathrm{d}x = B \right\}, \tag{8}$$

$$h(\lambda,\nu) := \inf\left\{ \int_{S(u)\cap Q_\nu} \Psi([u],\nu(u))\,\mathrm{d}\mathcal{H}^{N-1} : \right.$$
$$\left. u \in SBV(Q_\nu;\mathbb{R}^d),\ u|_{\partial Q_\nu} = u_{\lambda,\nu}(x),\ \nabla u = 0\ \mathcal{L}^N\text{-a.e.} \right\}, \tag{9}$$

with $u_{\lambda,\nu}(x) := \begin{cases} 0 & \textit{if } -\frac{1}{2} \le x\cdot\nu < 0, \\ \lambda & \textit{if } 0 \le x\cdot\nu < \frac{1}{2}. \end{cases}$

The relaxation described in Theorem 3 is based on the so-called *blow-up method* [15]: the energy densities H and h are obtained as the energetically optimal ones realizing an affine transformation (for the bulk part) with prescribed boundary values and gradient, and a jump (for the surface part) with prescribed jump height. This is essentially the meaning of the conditions in the infimization problems (8) and (9).

Results analogous to that stated in Theorem 3 are available in the literature: the reader is referred to [8] for the case $p = 1$, and to [5, 19, 21] for second-order structured deformations and for different types of initial energies (6).

3 Three Problems

3.1 Explicit Formulas for Purely Interfacial Energies

In the case $d = N$, we investigate an energy functional of the type (6) with $W = 0$, that is, where the energy is purely interfacial. By applying the blow-up method, we obtain a representation formula of the relaxed energy (7) for particular choices of initial surface energy densities, namely for $E^{|\cdot|}(u) := \int_{S(u)\cap\Omega} |[u]\cdot\nu(u)|\,\mathrm{d}\mathcal{H}^{N-1}$ and $E^{\pm}(u) := \int_{S(u)\cap\Omega} ([u]\cdot\nu(u))^{\pm}\,\mathrm{d}\mathcal{H}^{N-1}$. Given $(g,G) \in SD(\Omega)$, denote by $\mathcal{V}^{|\cdot|}(g,G)$ and $\mathcal{V}^{\pm}(g,G)$ the relaxed energies obtained via Theorem 3.

Theorem 4 (see [4, 20, 23]) *The initial disarrangement densities* $\Psi^{|\cdot|}(\lambda,\nu) := |\lambda\cdot\nu|$ *and* $\Psi^{\pm}(\lambda,\nu) := (\lambda\cdot\nu)^{\pm}$ *have relaxed disarrangement densities given by* $H^{|\cdot|}(A,B) = |\operatorname{tr}(A-B)|$, $h^{|\cdot|}(\lambda,\nu) = \Psi^{|\cdot|}(\lambda,\nu)$, *and* $H^{\pm}(A,B) = (tr(A-B))^{\pm}$, $h^{\pm}(\lambda,\nu) = \Psi^{\pm}(\lambda,\nu)$.

The proof presented in [4] that $H^{|\cdot|}(A,B) = |\operatorname{tr}(A-B)|$ relies on the following chain of inequalities

$$
\begin{aligned}
|\mathrm{tr}(A-B)| &\le \inf\left\{\int_{S(u)\cap Q}|[u](x)\cdot\nu(u)(x)|\,\mathrm{d}\mathscr{H}^{N-1}(x):\ u\in SBV(Q;\mathbb{R}^N),\right.\\
&\qquad \left. u|_{\partial Q}=Ax,\ \nabla u\in L^p(Q),\ \int_Q \nabla u\,\mathrm{d}x = B\right\}\\
&\le \inf\left\{\int_{S(u)\cap Q}|[u](x)\cdot\nu(u)(x)|\,\mathrm{d}\mathscr{H}^{N-1}(x):\ u\in SBV(Q;\mathbb{R}^N),\right.\\
&\qquad \left. u|_{\partial Q}=0,\ \nabla u = B-A \text{ a.e. in } Q\right\} \le |\,\mathrm{tr}(A-B)|.
\end{aligned}
\tag{10}
$$

The first inequality comes from (8), the second one holds because we are restricting the set of admissible functions in the minimization, the third one is proved in [4]. It is worth noticing that the second minimization problem in (10) is the one proposed in [3] in a different context for relaxation; the proof of the third inequality in (10) is achieved by an explicit construction and on the notion of *isotropic vectors* [9].

Other than recovering the explicit results of [20] with a shorter proof, and other additional explicit formulas, (10) shows the equivalence of two minimum problems. The recent paper [23] contains results on the general form of the relaxation of purely interfacial energies, which extends the previous results.

Another consequence is the relationship between the relaxed energies for the three densities considered in Theorem 4, namely (recalling (5))

$$
\mathscr{V}^{\pm}(g,G)=\frac{1}{2}\mathscr{V}^{|\cdot|}(g,G)\pm\frac{1}{2}\int_\Omega \mathrm{tr}\, M(x)\,\mathrm{d}x.
$$

3.2 Optimal Design

In the context of optimal design, we consider a two-component fractured medium with prescribed macroscopic strain. The initial energy functional of type (6) is tailored to account for the fine structure of the material and the relaxed energy densities are obtained by the interplay between the optimization of sharp interfaces and the diffusion of microscopic cracks.

Let $\chi\in BV(\Omega;\{0,1\})$ be the characteristic function of a set of finite perimeter (see [2]) describing one constituent of the material. Given a deformation $u\in SBV(\Omega;\mathbb{R}^d)$, consider the initial energy

$$
\begin{aligned}
E(\chi,u) &:= \int_\Omega ((1-\chi)W^0(\nabla u)+\chi W^1(\nabla u))\,\mathrm{d}x\\
&+\int_{\{\chi=0\}\cap S(u)\cap\Omega}\Psi_1^0([u],\nu(u))\,\mathrm{d}\mathscr{H}^{N-1}+\int_{\{\chi=1\}\cap S(u)\cap\Omega}\Psi_1^1([u],\nu(u))\,\mathrm{d}\mathscr{H}^{N-1}\\
&+\int_{S(\chi)\cap S(u)\cap\Omega}\Psi_2(\chi^+,\chi^-,u^+,u^-,\nu(u))\,\mathrm{d}\mathscr{H}^{N-1}+|D\chi|(\Omega),
\end{aligned}
$$

which features the contributions of bulk and surface energy densities W^i and Ψ_1^i, $i = 0, 1$, for the deformation u, as well as a surface energy density Ψ_2 on the superposition of the singular sets of u and χ, and a perimeter penalization term pushing for smaller interfaces between the two constituents.

We assume that the energy densities W^i satisfy (H_1) and are coercive, that Ψ_1^i satisfy (H_2)-(H_4), and that Ψ_2 satisfies

(H_5) $\exists\, C > 0$ such that $\forall\, a, b \in \{0, 1\}$, $c, d \in \mathbb{R}^d$, $\nu \in \mathbb{S}N-1$: $0 \le \Psi_2(a, b, c, d, \nu) \le C(1 + |a-b| + |c-d|)$;

(H_6) $\forall\, a, b \in \{0, 1\}, c, d \in \mathbb{R}^d, \nu \in \mathbb{S}N-1$: $\Psi_2(a, b, c, d, \nu) = \Psi_2(b, a, d, c, -\nu)$;

(H_7) $\exists\, C > 0$ such that $\forall\, a, b, \in \{0, 1\}$, $c_i, d_i \in \mathbb{R}^d$, $i = 1, 2$, $\nu \in \mathbb{S}N-1$, $|\Psi_2(a, b, c_1, d_1, \nu) - \Psi_2(a, b, c_2, d_2, \nu)| \le C|(c_1 - d_1) - (c_2 - d_2)|$;

(H_8) $\forall\, a \in \{0, 1\}$, $c \in \mathbb{R}^d$: $\Psi_2(\cdot, \cdot, c, c, \cdot) = \Psi_2(a, a, \cdot, \cdot, \cdot) = 0$.

Given $(\chi, g, G) \in BV(\Omega; \{0, 1\}) \times SD(\Omega)$, the relaxed energy is defined by

$$I(\chi, u, G) := \inf\Big\{ \liminf_{n\to\infty} E(\chi_n, u_n) : \chi_n \in BV(\Omega; \{0, 1\}),\ u_n \in SBV(\Omega; \mathbb{R}^d),$$
$$\chi_n \overset{*}{\rightharpoonup} \chi \text{ in } BV(\Omega; \{0, 1\}),\ u_n \to u \text{ in } L^1(\Omega; \mathbb{R}^d),\ \nabla u_n \rightharpoonup G \text{ in } L^p(\Omega; \mathbb{R}^{d\times N}) \Big\},$$

and the relaxation theorem states the following.

Theorem 5 ([16, Theorem 3.3]) *Let $(\chi, g, G) \in BV(\Omega; \{0, 1\}) \times SD(\Omega)$ and let the bulk and surface energy densities W^i, Ψ_1^i ($i = 0, 1$), and Ψ_2 satisfy (H_1)-(H_8). Then*

$$I(\chi, g, G) = \int_\Omega H(\chi, \nabla g, G)\,\mathrm{d}x + \int_{S(\chi,g)\cap\Omega} h(\chi^+, \chi^-, g^+, g^-, \nu)\,\mathrm{d}\mathscr{H}^{N-1}$$

where, for $i, a, b \in \{0, 1\}$, $A, B \in \mathbb{R}^{d\times N}$, $c, d \in \mathbb{R}^d$, and $\nu \in \mathbb{S}N-1$,

$$H(i, A, B) := \inf\Big\{ \int_Q W^i(\nabla u)\,\mathrm{d}x + \int_{S(u)\cap Q} \Psi_1^i([u], \nu(u))\,\mathrm{d}\mathscr{H}^{N-1} :$$
$$u \in SBV(Q; \mathbb{R}^d),\ |\nabla u| \in L^p(Q),\ u|_{\partial Q} = Ax,\ \int_Q \nabla u\,\mathrm{d}x = B \Big\},$$

$$h(a, b, c, d, \nu) := \inf\Big\{ \int_{S(\chi)\cap S(u)\cap Q_\nu} \Psi_2(\chi^+, \chi^-, u^+, u^-, \nu(u))\,\mathrm{d}\mathscr{H}^{N-1} + |D\chi|(Q_\nu)$$
$$+ \int_{\{\chi=0\}\cap S(u)\cap Q_\nu} \Psi_1^0([u], \nu(u))\,\mathrm{d}\mathscr{H}^{N-1}$$
$$+ \int_{\{\chi=1\}\cap S(u)\cap Q_\nu} \Psi_1^1([u], \nu(u))\,\mathrm{d}\mathscr{H}^{N-1} : (\chi, u) \in \mathscr{A}(a, b, c, d, \nu) \Big\},$$

where $\mathscr{A}(a,b,c,d,\nu) := \{(\chi,u) \in BV(Q_\nu;\{0,1\}) \times SBV(Q_\nu;\mathbb{R}^d) : \chi|_{\partial Q_\nu} = \chi_{a,b,\nu},\ u|_{\partial Q_\nu} = u_{c,d,\nu},\ \nabla u = 0\ \mathscr{L}^N\text{-a.e.}\}$ *with*

$$\chi_{a,b,\nu}(x) := \begin{cases} a & \text{if } x\cdot\nu > 0,\\ b & \text{if } x\cdot\nu \le 0, \end{cases} \qquad \text{and} \quad u_{c,d,\nu}(x) := \begin{cases} c & \text{if } x\cdot\nu > 0,\\ d & \text{if } x\cdot\nu \le 0. \end{cases}$$

3.3 Dimension Reduction

Dimension reduction is a technique to study thin objects in 2d starting from a fully 3d model and energetics. A small thickness parameter $\varepsilon > 0$ is involved in the 3d energy and sent to zero to obtain the 2d energy. We consider a bounded open set $\omega \subset \mathbb{R}^2$ and define $\Omega_\varepsilon := \omega \times (-\frac{\varepsilon}{2}, \frac{\varepsilon}{2})$; for $u \in SBV(\Omega_\varepsilon;\mathbb{R}^3)$ let

$$E_\varepsilon(u) := \int_{\Omega_\varepsilon} W_{3d}(\nabla u(x))\,\mathrm{d}x + \int_{S(u)\cap\Omega_\varepsilon} \Psi_{3d}([u](x),\nu(u)(x))\,\mathrm{d}\mathscr{H}^2,$$

where the densities W_{3d} and Ψ_{3d} satisfy assumptions (H_1)-(H_4).

In order to couple dimension reduction and structured deformations, we shall perform the two relaxation processes one after the other in two different orders, as depicted in the left- and right-hand side paths in Fig. 2. The dimension reduction legs DR are obtained by the change of variables $x_3 \to x_3/\varepsilon$, so that the functional will be defined on the volume $\Omega := \Omega_1 = \omega \times (-\frac{1}{2}, \frac{1}{2})$. After rescaling the functional by dividing it by ε, the limit as $\varepsilon \to 0$ is computed. The structured deformation legs SD are obtained by applying the relaxation Theorem 3. It is important to point out that, in the dimension reduction process, there is the emergence of a vector $\overline{d} : \omega \to \mathbb{R}^3$ that keeps memory of the behavior of the deformation in the x_3-direction. The role played by $\overline{d}$ is encoded via the constraint $\int_{-1/2}^{1/2} \frac{\nabla_3 u_n}{\varepsilon_n}\,\mathrm{d}x_3 \rightharpoonup \overline{d}$ in the relaxation process (see [6]).

A comparison of the relaxed energy densities is interesting, to see whether or not the two different ways provide the same result. In this respect, partial results are available in the special case of the purely interfacial initial energy densities in

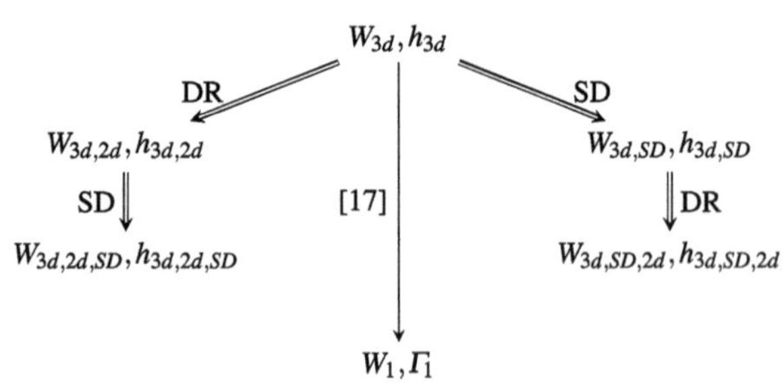

Fig. 2 Energy densities for the paths for dimension reduction and structured deformations in [6]

Theorem 4. Moreover, in this case, another relaxation procedure is available in the literature: in [17] a relaxation that simultaneously defines a 2d energy on structured deformations is studied. We prove [6, Sect. 5] that the left- and right-hand paths provide the same relaxed energy densities, whereas those computed using the central path in Fig. 2 are lower.

4 Conclusions and Outlook

In this brief note, we have presented a general overview of structured deformations and their versatility by describing three applications which are very different from one another. In Sect. 3.1 we obtained explicit formulas and the equivalence of two minimum problems [4]; in Sect. 3.2 we coupled structured deformations to study an optimal design problem, and performed a relaxation in the joint variables χ and u [16]; finally, in Sect. 3.3 we coupled structured deformations with dimension reduction, showing two possible routes to obtain the final relaxed energies, and comparing the results for a specific choice of initial energy densities [6].

The results in Sect. 3.1 can be extended to more general situations and call for the search of explicit formulas for other initial energy densities: these will be certainly useful in the applications. Those in Sect. 3.2 are a first step towards the far-reaching goal of incorporating elements of plasticity in optimal design of composite media. Section 3.3 presents challenges at a more theoretical level, regarding the commutativity of different relaxation processes and the comparison of minimum problems for the same functional, but different sets of admissible functions.

The versatility of structured deformations makes them suitable to model physical systems such as composite or granular media, defective materials, and biological membranes with the aim of shedding new light on the mechanics of these systems.

Acknowledgements The author acknowledges partial support for this research from the following grants: FCT_UTA/CMU/MAT/0005/2009 of the Fundação para a Ciência e a Tecnologia through the Carnegie Mellon Portugal Program; ERC Advanced grant *Quasistatic and Dynamic Evolution Problems in Plasticity and Fracture* (Grant agreement 290888); INdAM-GNAMPA project 2015 *Fenomeni Critici nella Meccanica dei Materiali: un Approccio Variazionale*; ERC Starting grant *High-Dimensional Sparse Optimal Control* (Grant agreement 306274). The author is a member of the GNAMPA group of INdAM. The author is grateful to J. Matias and D. R. Owen for valuable suggestions in writing this note.

References

1. Alberti, G.: J. Funct. Anal. **100**, 110–118 (1991)
2. Ambrosio, L., Fusco, N., Pallara, D.: OUP, (2000). ISBN: 9780198502456
3. Baía, M., Matias, J., Santos, P.M.: Proc. Royal Soc. Edinb. **142A**, 239–271 (2012)
4. Barroso, A.C., Matias, J., Morandotti, M., Owen, D.R.: Math. Mech. Complex Syst. **5**(2), 163–189 (2017)

5. Barroso, A.C., Matias, J., Morandotti, M., Owen, D.R.: Arch. Rational Mech. Anal. **225**(3), 1025–1072 (2017)
6. Carita, G., Matias, J., Morandotti, M., Owen, D.R.: Dimension reduction in the context of structured deformations (2017). arXiv: 1709.02869
7. Choksi, R., Del Piero, G., Fonseca, I., Owen, D.R.: Math. Mech. Solids. **4**, 321–356 (1999)
8. Choksi, R., Fonseca, I.: Arch. Rational Mech. Anal. **138**, 37–103 (1997)
9. Ciblak, N., Lipkin, H.: In: Proceedings of DETC'98, Atlanta, Georgia Sep 13–16 (1998)
10. Del Piero, G., Owen, D.R.: Arch. Ration. Mech. Anal. **124**, 99–155 (1993)
11. Del Piero, G., Truskinovsky, L.: Int. J. Solids Struct. **38**, 1135–1148 (2001)
12. Deseri, L., Owen, D.R.: J. Elast. **70**, 197–236 (2003)
13. Deseri, L., Owen, D.R.: Int. J. Eng. Sci. **96**, 111–140 (2015)
14. Fonseca, I., Leoni, G.: Springer, New York, (2007). ISBN 978-0-387-69006-3
15. Fonseca, I., Müller, S.: Arch. Ration. Mech. Anal. **123**, 1–49 (1993)
16. Matias, J., Morandotti, M., Zappale, E.: J. Math. Anal. Appl. **449**, 1094–1132 (2017)
17. Matias, J., Santos, P.M.: Appl. Math. Optim. **69**, 459–485 (2014)
18. Owen, D.R.: J. Elasticity **127**(1), 115–150 (2017)
19. Owen, D.R., Paroni, R.: Arch. Ration. Mech. Anal. **155**, 215–235 (2000)
20. Owen, D.R., Paroni, R.: Arch. Ration. Mech. Anal. **218**, 1633–1652 (2015)
21. Paroni, R.: In: Del Piero, G., Owen, D.R. (eds.) CISM. vol. 447, Springer (2004)
22. Šilhavý, M.: Math. Mech. Complex Syst. **3**, 83–100 (2015)
23. Šilhavý, M.: Math. Mech. Complex Syst. **5**(2), 191–215 (2017)

Part V
Interface Dynamics

Gradient Flows with Wiggly Potential: A Variational Approach to Dynamics

Nadia Ansini

Abstract Free energies with many small wiggles, arising from small-scale micro-structural changes, appear often in phase transformations, protein folding and friction problems. In this paper, we investigate gradient flows with energies E_ε given by the superposition of a convex functional and fast small oscillations. We apply the time-discrete minimising-movement scheme to capture the effect of the local minimisers of E_ε in the limit equation as ε tends to zero. We perform a multiscale analysis according to the mutual vanishing behaviour of the spatial parameter ε and the time step τ, and we highlight three different regimes $\tau \ll \varepsilon$, $\varepsilon \ll \tau$ (Braides, Local Minimization, Variational Evolution and Γ-convergence. Springer, Cham (2014), [4]) and $\tau \sim \varepsilon$ (Ansini et al., Minimising movements for oscillating energies: the critical regime, [3]). We discuss for each case the existence of a pinning threshold, and we derive the limit equation describing the motion.

1 Introduction

In this paper, we analyse a minimising-movement approach for gradient flows with wiggly energies,

$$x'(t) = -\mu \frac{\partial E_\varepsilon(x(t))}{\partial x}. \tag{1}$$

A prototypical model of the energy is an oscillating perturbation of a quadratic energy,

$$E_\varepsilon(x) = \frac{1}{2}x^2 - \varepsilon \cos\left(\frac{x}{\varepsilon}\right). \tag{2}$$

Free energies with many small wiggles, arising from small-scale micro-structural changes, appear often in phase transformations, protein folding and friction problems.

N. Ansini (✉)
Department of Mathematics, Sapienza University of Rome, P.le Aldo Moro 2, 00185 Rome, Italy
e-mail: ansini@mat.uniroma1.it

P. van Meurs et al. (eds.), *Mathematical Analysis of Continuum Mechanics and Industrial Applications II*, Mathematics for Industry 30,
https://doi.org/10.1007/978-981-10-6283-4_12

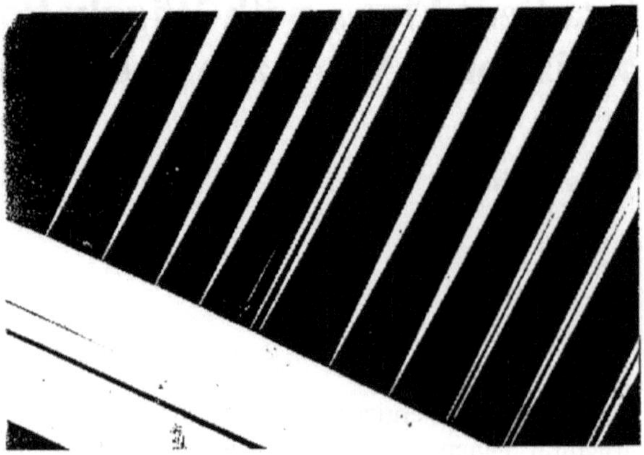

Fig. 1 Picture of tip splitting: a band in the centre has just split. Reprinted by permission of Taylor & Francis Ltd from [1]

In particular, in materials undergoing martensitic phase transitions the formation of needles of phases with pronounced tips has been observed in experiments on the transition between two variants of martensite in single crystals (see [1]). During creep tests, it is observed that the volume fraction of the phase fractions involved changes rather abruptly, and it is shown that this sudden change can be attributed to a sudden split of a tip into two tips. One can picture this as a lenticular domain of one variant trying to grow; this growth then occurs where the tip of the lens meets a boundary between twins, and fattening of the phase happens via splitting of the tip in two and more tips (see Fig. 1).

The splitting of a needle can then be attributed to a metastable transition, moving from one local minimum to another one. This suggests a small-scale landscape with many minima, and the energy studied by Abeyaratne, Chu and James [1] is a macroscopic energy augmented by small-scale oscillations $a\varepsilon\cos(\frac{x}{\varepsilon})$, as studied here. In addition, the kinetic law in [1] is taken to be a gradient flow. Specifically, there it is shown that the solution x_ε to the evolution Eq. (1) converges uniformly in time to the solution of

$$x'(t) = -\mu \frac{\partial \bar{E}(x(t))}{\partial x}, \text{ with } x(0) = x_0, \tag{3}$$

with an explicitly computed driving force $\frac{\partial \bar{E}}{\partial x}$. This latter system is then investigated numerically. The variational analysis carried out here can be interpreted in this light. We consider time discretisations, as numerical algorithms would employ, but on the level of the original (not homogenised) energy E_ε, rather than $\bar{E}$. This leads to two parameters, the time discretisation τ and oscillation scale ε. The different scaling regimes that follow naturally are analysed in [3, 4].

The aim of the paper is to provide an overview of the multiscale analysis performed according to the mutual vanishing behaviour of the parameters ε and τ. We summarise motivations, variational techniques and results obtained.

In particular, for $\tau \ll \varepsilon$ (*fast time scale*), we show that the minimising movement x coincides with the limit of the solutions x_ε of the gradient flows at fixed ε, while for $\varepsilon \ll \tau$ (*slow time scale*) it coincides with the gradient flow of the limit quadratic energy that is trivially the $\Gamma\text{-}\lim_{\varepsilon\to 0} E_\varepsilon(x) = -(x^2/2)$ (see [4]). That observation highlights the existence of a *critical time scaling* when $\tau \sim \varepsilon$, for which the minimising movements are not trivially described by the limit of gradient flows or the gradient flow of the limit. The behaviour at those scales is the object of the paper [3] where the authors consider the following energies

$$E_\varepsilon(x) = h(x) + \varepsilon\, W\Big(\frac{x}{\varepsilon}\Big), \tag{4}$$

where h is a strictly convex function bounded from below and W is a one-periodic even Lipschitz function.

We remark that the study of gradient flows with wiggly energies allows us to explore different phenomena that may appear in cases where we do not have stability of the minimising-movement scheme with respect to the Γ-convergence. Indeed, the energies E_ε, that we consider in this paper, do not satisfy the conditions that ensure that all minimising movements along a family of energies, at whatever scale, converge to minimising movements of their Γ- limit (see e.g. [6] and references therein).

2 Minimising Movement Along a Sequence

We recall the general definition of minimising movements for a sequence of functionals defined on a Hilbert space.

Definition 1 Let X be a separable Hilbert space, $E_\varepsilon : X \to [0, +\infty)$ equicoercive and lower semicontinuous and $x_0^\varepsilon \to x_0$ with $E_\varepsilon(x_0^\varepsilon) \le C < +\infty$ and $\tau_\varepsilon > 0$ converging to 0 as $\varepsilon \to 0$. For fixed $\varepsilon > 0$, we define recursively x_i^ε as a minimiser of the problem

$$\min\Big\{E_\varepsilon(x) + \frac{1}{\tau_\varepsilon}\|x - x_{i-1}^\varepsilon\|^2\Big\}, \tag{5}$$

and the piecewise-constant trajectory

$$x^\varepsilon(t) := x_i^\varepsilon = x^\varepsilon_{\lfloor t/\tau_\varepsilon \rfloor}, \quad t \in [i\tau_\varepsilon, (i+1)\tau_\varepsilon)\,. \tag{6}$$

A *minimising movement for* E_ε *at time scale* τ from x_0^ε is the limit of a subsequence x^{ε_j},

$$x(t) = \lim_{j\to+\infty} x^{\varepsilon_j}(t)\,, \tag{7}$$

with respect to the uniform convergence on compact sets of $[0, +\infty)$.

This definition is justified by the following compactness result [4, Proposition 7.1].

Proposition 1 *For every E_ε and x_0^ε as above, there exist minimising movements for E_ε, from x_0^ε, with $x(t) \in C^{1/2}([0, +\infty); X)$.*

For a comprehensive study of minimising movements for a fixed $E = E_\varepsilon$, we refer to [2] (this notion coincides with the one given by De Giorgi [7]), while a detailed analysis of some of its applications can be found in [4].

3 Monotone Behaviour of Minimising Movements

In the sequel, we will study minimising movements for the functions $E_\varepsilon \colon \mathbf{R} \to \mathbf{R}$ given by (4), where h is a strictly convex function bounded from below. It is not restrictive to suppose that $h \geq 0$, and that h attains its global minimum in $x = 0$. Furthermore, we assume that W is a one-periodic even Lipschitz function with $\|W'\|_\infty = 1$, and that the average of W is 0. The two latter assumptions serve as normalisation only and are not restrictive.

We start by making some general remarks. According to Definition 1, we define iteratively the global minimiser x_{i+1}^ε to

$$F_\varepsilon(x, x_i^\varepsilon) = h(x) + \varepsilon\, W\Big(\frac{x}{\varepsilon}\Big) + \frac{1}{2\tau_\varepsilon}(x - x_i^\varepsilon)^2\,. \tag{8}$$

First of all, we observe that the sequence of minimisers $(x_i^\varepsilon)_i$ is monotone.

Proposition 2 (Monotone behaviour of x_i^ε) *Let x_{i+1}^ε be a minimiser to (8). Then, the following holds.*

1. *If $x_{i+1}^\varepsilon \leq x_i^\varepsilon$, then $x_{i+2}^\varepsilon \leq x_{i+1}^\varepsilon$.*
2. *If $x_{i+1}^\varepsilon \geq x_i^\varepsilon$, then $x_{i+2}^\varepsilon \geq x_{i+1}^\varepsilon$.*

In particular, $t \mapsto x^\varepsilon(t)$ and $t \mapsto x(t)$ are monotone functions.

In order to characterise the velocity $x'(t_0)$ of the minimising-movement scheme (5), we study the average velocity given by

$$\frac{x_i^\varepsilon - x_0^\varepsilon}{i\tau_\varepsilon}\,. \tag{9}$$

We assume, without loss of generality, that $t_0 = 0$ and $x(0) = x_0^\varepsilon$. We consider a (partial) linearisation of the problem given by

$$F_\varepsilon^T(x, x_i^{\varepsilon,T}) = Tx + \varepsilon\, W\Big(\frac{x}{\varepsilon}\Big) + \frac{1}{2\tau_\varepsilon}(x - x_i^{\varepsilon,T})^2\,, \tag{10}$$

where $x_0^{\varepsilon,T} := x_0^\varepsilon$. The term Tx represents the "linear approximation" of the potential h around the point $h(x_0^\varepsilon)$ up to translation by a constant that does not depend on T

and i. We recall that h is a strictly convex function, hence h' is a monotone increasing function.

Proposition 3 (Monotone behaviour of $x_i^{\varepsilon,T}$ with respect to x_i^{ε}) *Let $\delta > 0$ be such that $h'(x_0 \pm \delta)$ exists.*

1. *Let $x_i^{\varepsilon,T}$ be the minimiser to (10). Then,*
   - *if $x_{i+1}^{\varepsilon,T} \le x_i^{\varepsilon,T}$, then $x_{i+2}^{\varepsilon,T} \le x_{i+1}^{\varepsilon,T}$.*
   - *if $x_{i+1}^{\varepsilon,T} \ge x_i^{\varepsilon,T}$, then $x_{i+2}^{\varepsilon,T} \ge x_{i+1}^{\varepsilon,T}$.*
2. *Let $T = T^{\delta+} := h'(x_0 + \delta)$, and x_i^{ε} minimiser of (8). Then if $x_i^{\varepsilon,T^{\delta+}} \le x_i^{\varepsilon}$, then $x_{i+1}^{\varepsilon,T^{\delta+}} \le x_{i+1}^{\varepsilon}$.*
3. *Let $T = T^{\delta-} := h'(x_0 - \delta)$; then if $x_i^{\varepsilon} \le x_i^{\varepsilon,T^{\delta-}}$, then $x_{i+1}^{\varepsilon} \le x_{i+1}^{\varepsilon,T^{\delta-}}$.*

4 Multiscale Analysis

A minimising movement $x(t)$ along the sequence of energies E_ε at time scale τ, with fixed $\tau = \tau(\varepsilon)$, is defined as any limit of subsequences of $x^{\varepsilon,\tau}(t) = x^{\varepsilon,\tau}_{\lfloor t/\tau \rfloor}$. Simple examples show that the limit may indeed depend on the subsequence and on the choice of τ. Therefore, a multiscale analysis has to be performed in order to determine the limit gradient flow at different regimes.

We consider a time scale $\tau = \tau(\varepsilon)$ such that ε/τ converges to γ. In Sect. 4.1, we study the cases $\gamma = 0, +\infty$ investigated in [4] (see also [5]).

In Sect. 4.2, we complement the previous analysis by considering $\gamma > 0$ as investigated in [3]. In terms of the mechanical problem of interface propagation discussed in Sect. 1, we show that pinning will occur for small initial data, while large data leads to a gradient flow evolution for which the averaged velocity can be computed. More precisely, we prove that there exists a pinning threshold T_γ such that the unique minimising movement x^γ with initial datum $|x_0| \le T_\gamma$ is $x^\gamma(t) = x_0$ for all t while if $x_0 > T_\gamma$ then $x^\gamma(t)$ is characterised as the non-increasing function satisfying

$$\frac{\mathrm{d}}{\mathrm{dt}} x^\gamma(t) = -\gamma\, f_\gamma(h'(x^\gamma(t)))$$

at almost all $t > 0$. The *homogenised velocity* $f_\gamma(z)$ is the *average velocity* (suitably defined) of any discrete orbit $\{y_k\}$ defined recursively by minimisation of the linearity problem

$$y \mapsto zy + W(y) + \frac{\gamma}{2}(y - y_{k-1})^2, \tag{11}$$

which can be shown not to depend on the initial condition y_0. The case $x_0 < 0$ is analogous by symmetry.

Mathematically, our analysis is confined to one space dimension, as it strongly relies on monotonicity properties developed in Sect. 3. A central argument is a comparison of solution to the nonlinear energy as in (4), and one where h, the non-oscillating part, is suitably linearised.

4.1 The Extreme Cases: Fast Time Scales and Slow Time Scales

For the energy (2), the associated gradient flow is ($\mu = 1$)

$$x_\varepsilon'(t) = -x_\varepsilon(t) - \sin\left(\frac{x_\varepsilon(t)}{\varepsilon}\right), \tag{12}$$

for initial datum x_0.

If $\tau \ll \varepsilon$ (*fast time scale*), the minimising movement x coincides with the limit of the solutions x_ε of the gradient flows at fixed ε. Indeed, if $x_0 \in (-1, 1)$, then such solutions are trapped between stationary solutions, and they converge to the trivial constant state x_0 (*pinning*) (see Figs. 2 and 3), while if $|x_0| \geq 1$ they can be shown to converge to a solution x of the gradient flow

$$x'(t) = -\operatorname{sign} x(t)\sqrt{x^2(t) - 1}. \tag{13}$$

On the other hand, E_ε converge uniformly to the quadratic energy, whose gradient flow is trivially

$$x'(t) = -x(t). \tag{14}$$

If $\varepsilon \ll \tau$ (*slow time scale*), the minimising movement x coincides with the solution of the gradient flow of the limit quadratic energy (see Fig. 3).

If E_ε is given by (4), we have that the equations

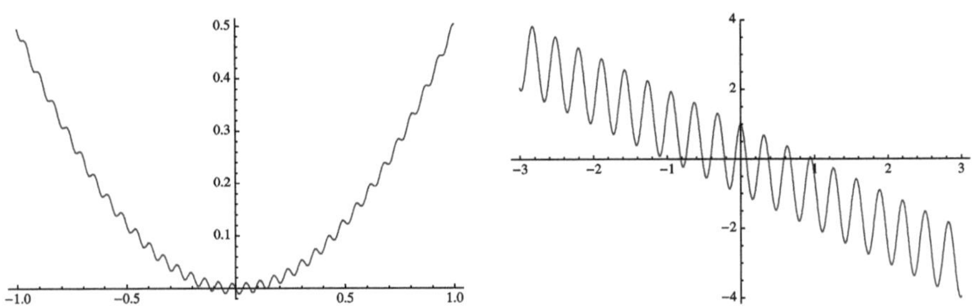

Fig. 2 Graphs of E_ε, E_ε'

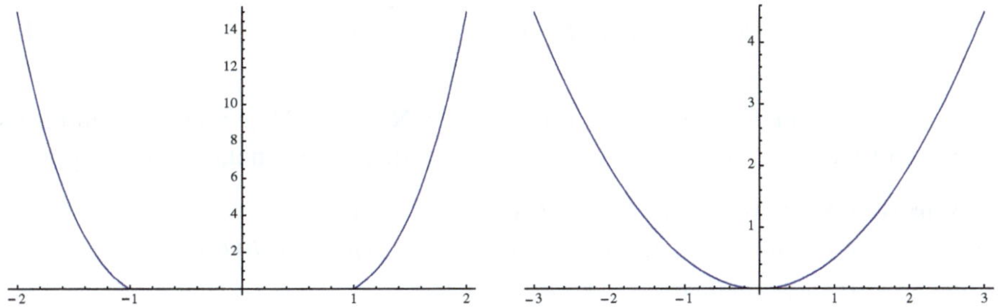

Fig. 3 Graphs of limit energies when $\tau \ll \varepsilon$ and $\varepsilon \ll \tau$

$$x' = -h'(x) \text{ and } x' = -g_\infty(h'(x)) \tag{15}$$

describe the minimising movements in the cases $\varepsilon \ll \tau$ and $\tau \ll \varepsilon$, respectively, where g_∞ is given by

$$g_\infty(z) = \begin{cases} \left(\int_0^1 \frac{1}{z + W(s)} ds\right)^{-1} & \text{if } \frac{1}{z+W(s)} \text{is integrable,} \\ 0 & \text{otherwise} \end{cases} \tag{16}$$

(see [4]). The results obtained in the cases $\varepsilon \ll \tau$ and $\tau \ll \varepsilon$ highlight the existence of a *critical time scaling* when $\tau \sim \varepsilon$, for which the minimising movements are not trivially described by the limit of gradient flows or the gradient flow of the limit. The behaviour at those scales is the object of the paper [3].

4.2 The Critical Case: Minimising Movement for Fixed Ratio ε/τ

We consider a time scale $\tau = \tau_\varepsilon$ such that ε/τ converges to $\gamma > 0$. It is not restrictive to suppose that the ratio between ε and τ is fixed,

$$\tau_\varepsilon = \frac{1}{\gamma}\varepsilon. \tag{17}$$

We study the linearised energies in (10) and rescale by $1/\varepsilon$, that is,

$$\frac{F_\varepsilon^T(x, x_i^{\varepsilon,T})}{\varepsilon} = T\Big(\frac{x}{\varepsilon}\Big) + W\Big(\frac{x}{\varepsilon}\Big) + \frac{\varepsilon}{2\tau_\varepsilon}\Big(\frac{x - x_i^{\varepsilon,T}}{\varepsilon}\Big)^2. \tag{18}$$

We denote

$$F_\gamma^T(y, y_i^T) := Ty + W(y) + \frac{\gamma}{2}(y - y_i^T)^2\,, \tag{19}$$

where $y := x/\varepsilon$ and $y_i^T := x_i^{\varepsilon,T}/\varepsilon$ for every $i \in \mathbf{N}$, $i \geq 1$. Note that the minimisers y_i^T depend also on γ. However, we omit this dependence in the notation for simplicity.

Proposition 4 *Let y_0, z_0 and T, S be fixed with $T \leq S$. Let $y_0^T = y_0$, $z_0^S = z_0$. Let y_i^T and z_i^S be minimisers to $F_\gamma^T(y, y_{i-1}^T)$ and $F_\gamma^S(y, z_{i-1}^S)$, respectively, for every $i \in \mathbf{N}$ with $i \geq 1$. If $z_0 \leq y_0$, then $z_i^S \leq y_i^T$ for every i.*

Theorem 1 *For every T, the limit*

$$f_\gamma(T) := \lim_{i\to\infty} \frac{y_0 - y_i^T}{i} \tag{20}$$

exists and it is independent of y_0. Moreover, the function $T \mapsto f_\gamma(T)$ is monotone increasing.

The existence of the limit is a straightforward consequence of the subadditivity of the sequence (y_i^T).

The function $T \mapsto f_\gamma(T)$ is independent of y_0 since we can always rewrite for $k < i$

$$\frac{y_0 - y_i^T}{i} = \frac{y_0 - y_k^T + y_k^T - y_i^T}{i} = \frac{y_0 - y_k^T}{i} + \frac{y_k^T - y_i^T}{i - k}\frac{i - k}{i}\,. \tag{21}$$

Hence,

$$\lim_{i\to\infty} \frac{y_0 - y_i^T}{i} = \lim_{i\to\infty} \frac{y_k^T - y_i^T}{i - k}\,. \tag{22}$$

Finally, we remark that by definition (20) and Proposition 4 the function $T \mapsto f_\gamma(T)$ is monotone increasingthat isis if $T \leq S$, then $f_\gamma(T) \leq f_\gamma(S)$.

We now show that the limit trajectory x satisfies

$$x'(t) = -\gamma\, f_\gamma(h'(x(t))) \tag{23}$$

for almost all $t > 0$, with f_γ defined in (20). This equation fully characterises x given the initial datum x_0.

Theorem 2 *Let $\gamma \in (0, +\infty)$. Let t_0 be such that $x'(t_0)$ exists.*

Then,

$$\gamma\, f_\gamma(h'(x(t_0)^-)) \leq -x'(t_0) \leq \gamma\, f_\gamma(h'(x(t_0)^+))\,. \tag{24}$$

Sketch of the proof.

As consequence of the monotone behaviour of x_i^ε and $x_i^{\varepsilon,T}$ (see Proposition 3, 2.-3.), we have that

$$\frac{x_i^{\varepsilon,T^{\delta+}} - x_0}{i\tau_\varepsilon} \leq \frac{x_i^\varepsilon - x_0}{i\tau_\varepsilon} \leq \frac{x_i^{\varepsilon,T^{\delta-}} - x_0}{i\tau_\varepsilon}\,. \tag{25}$$

The averaged velocity, as in (9), is given by

$$\frac{x_i^{\varepsilon,T} - x_0}{i\tau_\varepsilon} = \gamma \frac{y_i^T - y_0}{i}; \tag{26}$$

with definition (20), it follows that

$$-\gamma f_\gamma(h'(x_0+\delta)) = \lim_{i\to\infty} \frac{x_i^{\varepsilon,T^{\delta+}} - x_0}{i\tau_\varepsilon}, \qquad -\gamma f_\gamma(h'(x_0-\delta)) = \lim_{i\to\infty} \frac{x_i^{\varepsilon,T^{\delta-}} - x_0}{i\tau_\varepsilon}. \tag{27}$$

Therefore, we conclude that

$$\gamma f_\gamma(h'(x_0^-)) \le -x'(0) \le \gamma f_\gamma(h'(x_0^+)). \tag{28}$$

□

The inequality (28) proves that Eq. (23) fully characterises x when $t \mapsto x(t)$ is strictly monotone, so that the set of t such that

$$\gamma f_\gamma(h'(x(t)^-)) \ne \gamma f_\gamma(h'(x(t)^+)) \tag{29}$$

is of zero (Lebesgue) measure. By the monotonicity of f_γ, if x is not strictly monotone then it is constant, so again (23) is satisfied.

We now characterise the *pinning set*, that is, the set of initial data for which $x(t) = x_0$ for all $t > 0$.

Definition 2 (*Pinning threshold*) For fixed $\gamma > 0$, we define the *pinning threshold* at scale γ as $T_\gamma := \sup\{T : f_\gamma(T) = 0\}$.

Remark 1 Note that f_γ is monotonically increasing, and thus $f_\gamma = 0$ on $[0, T_\gamma]$ (see Fig. 4). Hence, for all x_0 with $|x_0| \le T_\gamma$ the motion is pinnedthat isis $x(t) = x_0$ for all t.

In [3, Proposition 5.4], we give a criterion for the computation of the pinning threshold. Moreover, we observed that, under suitable assumptions, the function f_γ, for all $\gamma > 0$, has the following asymptotic behaviour at the pinning threshold

$$f_\gamma(T) \sim \frac{1}{\log(T - T_\gamma)} \tag{30}$$

as $T \to T_\gamma^+$. This will be shown for piecewise quadratic energies W in detail in the next section.

Fig. 4 Graph of f_γ

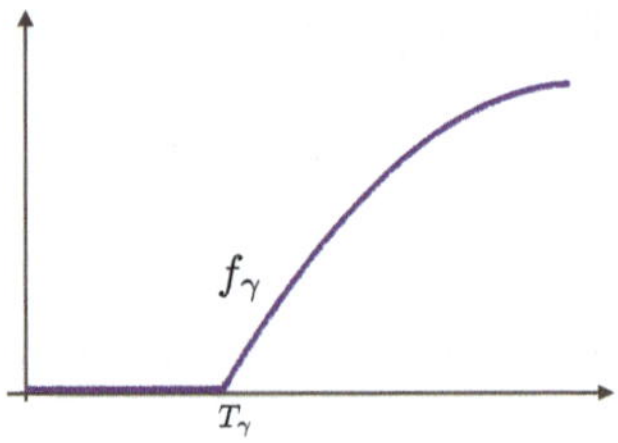

Proposition 5 (Extreme minimising movements) *We have*

$$\lim_{\gamma\to 0} \gamma f_\gamma(z) = z, \qquad \lim_{\gamma\to+\infty} \gamma f_\gamma(z) = g_\infty(z), \tag{31}$$

where g_∞ is given by (16). Moreover,

$$\lim_{\gamma\to+\infty} T_\gamma = \sup_\gamma T_\gamma = T_\infty, \tag{32}$$

where $[-T_\infty, T_\infty] = \{T \in \mathbf{R} : g_\infty(T) = 0\}$.

The previous proposition shows that the same extreme minimising movements are obtained by keeping the ratio $\gamma = \varepsilon/\tau$ fixed and then let it tend to 0 and $+\infty$, respectively (see Sect. 4.1).

5 An Example: The Piecewise Quadratic Case

In this section, we provide an example of oscillating potential and calculate explicitly the corresponding pinning threshold T_γ. More precisely, we consider the piecewise quadratic energy

$$W(y) := \min_{k\in\mathbf{Z}}(y-k)^2 . \tag{33}$$

Besides giving an illustrative example, we deduce the asymptotic behaviour at the pinning threshold, which depends only on the non-degeneracy of the second derivative at local minima. In this way, we deduce the asymptotic behaviour in the general case as in Sect. 4.2.

For this choice of W, for T_∞ as in Proposition 5, it holds that

$$T_\infty = 1, \tag{34}$$

and (16) becomes

$$g_\infty(z) = \frac{1}{\log\left(\frac{z-1}{z}\right)} \qquad \text{for } z > 1. \tag{35}$$

The function W is 1-periodic and piecewise quadratic with $\|W'\|_\infty = 1$. For simplicity, we fix also $h(x) = x^2/2$. Let $T \in (0, 1)$ and $y_0 \in [0, 1/2)$. The minimum of the function $F_\gamma^T(y, y_0) = Ty + W(y) + \frac{\gamma}{2}(y - y_0)^2$, given in (19), on the interval $[-1/2 + k, 1/2 + k]$ can be attained at the boundary or at the interior of this interval; it is given by

$$y_{1,k} = \frac{-T + 2k}{2 + \gamma} + \frac{\gamma}{2 + \gamma} y_0 \tag{36}$$

($y_{1,k}$ also depends on T but we suppress this in the notation).

The global minimiser y_1^T to $F_\gamma^T(y, y_0)$ can get stuck in the same well of y_0, that is, $y_1^T = y_{1,0}$. Otherwise, it can move into the next well, corresponding to $k = -1$, that is, $y_1^T = y_{1,-1}$.

Any single well of $W(y)$ is denoted by $W(y; k) := (y - k)^2$ for every $y \in [-1/2 + k, 1/2 + k]$ with $k \in \mathbf{Z}$. We define

$$\begin{aligned}\psi(y) := &\left(Ty_{1,-1} + W(y_{1,-1}; -1) + \frac{\gamma}{2}(y_{1,-1} - y)^2\right)\\ &-\left(Ty_{1,0} + W(y_{1,0}; 0) + \frac{\gamma}{2}(y_{1,0} - y)^2\right).\end{aligned} \tag{37}$$

To establish if y_1^T gets stuck or moves, we have to study the sign of $\psi(y_0)$, since it is the difference between the minimum value of the two wells. Therefore, if $\psi(y_0) < 0$, then the minimiser satisfies $y_1^T = y_{1,-1}$; $\psi(y_0) \geq 0$ implies that $y_1^T = y_{1,0}$. In particular, from the sign of $\psi(y_0)$ we expect to derive the pinning threshold T_γ.

Proposition 6 *Let $\gamma \in (0, +\infty)$ and let $T \in (0, 1)$. Then, there exist*

$$T_\gamma := \frac{\gamma}{(2 + \gamma)} \quad \text{and} \quad \delta_T := \left(\frac{2 + \gamma}{2\gamma}\right)(T - T_\gamma), \tag{38}$$

such that the following holds.

1. *For every $T > T_\gamma$ we have that, the following possibilities exist.*
   (i) *If $y_0 \in [0, (-T/2) + \delta_T)$, then $y_1^T = y_{1,-1}$, where the latter is defined in (36). Moreover, if*

$$y_1^T + \frac{T}{2} + 1 = \frac{\gamma}{1 + \gamma}\left(y_0 + \frac{T}{2} + 1\right) < -\frac{T}{2} + \delta_T, \tag{39}$$

   then the successive minimiser is given by $y_2^T = y_{2,-2}$ and so on.
   (ii) *If $y_0 \in [(-T/2) + \delta_T, 1/2)$, then there exists $h \in \mathbf{N}$ given by*

$$h = \left\lfloor \frac{\log\left(\left(\frac{2+\gamma}{\gamma}\right)\frac{T - T_\gamma}{T+1}\right)}{\log\left(\frac{\gamma}{2+\gamma}\right)} \right\rfloor + 1 \tag{40}$$

such that $y_1^T = y_{1,0}, \ldots, y_h^T = y_{h-1,0} \geq (-T/2) + \delta_T$*,* $y_{h,0} < (-T/2) + \delta_T$ *and* $y_{h+1}^T = y_{h+1,-1}$*. Similarly, if* $k \in \mathbf{N}$ *exists such that for some* $p \in \mathbf{N}$

$$y_k^T + \frac{T}{2} + p \geq -\frac{T}{2} + \delta_T \,, \tag{41}$$

then there exists h *as in (40) such that if we take as initial data* $z_0 := y_k^T + (T/2) + p$ *then we get a new sequence of minimisers such that* $z_1^T = z_{1,0}, \ldots, z_{h-1}^T = z_{h-1,0} \geq -T + \delta_T$*,* $z_h^T = z_{h,0} < (-T/2) + \delta_T$*, and* $z_{h+1}^T = z_{h+1,-1}$*.*

2 *For every* $T \leq T_\gamma$ *and* $y_0 \geq 0$*, we have that*

$$\lim_{h\to\infty} y_h^T = -\frac{T}{2}\,,$$

that is, the motion is pinned.

Remark 2 (*Behaviour at the pinning threshold*) From Proposition 6, case (ii), we deduce that

$$f_\gamma(T) \sim \frac{\log\left(\frac{\gamma}{2+\gamma}\right)}{\log\left(\left(\frac{2+\gamma}{\gamma}\right)\frac{T-T_\gamma}{T+1}\right)}; \tag{42}$$

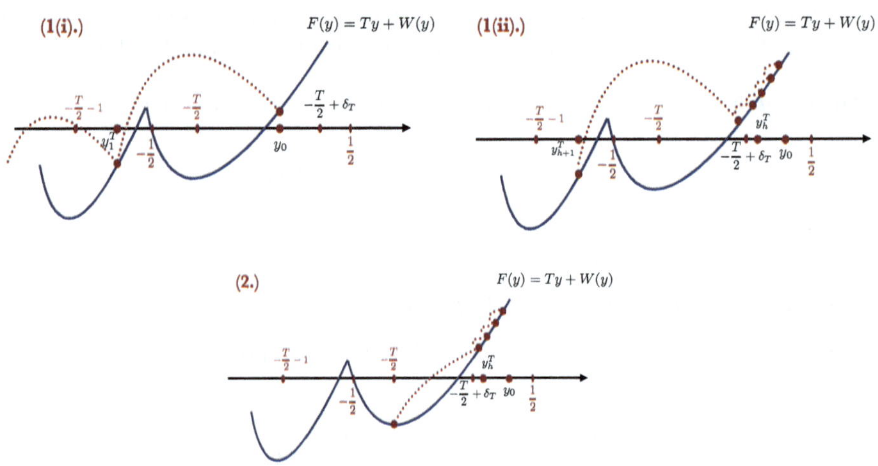

that is, for $\gamma > 0$ fixed,

$$f_\gamma(T) \sim \frac{1}{|\log(T - T_\gamma)|} \tag{43}$$

as $T \to T_\gamma^+$. Note in particular that f_γ is not Lipschitz for $T \to T_\gamma^+$.

Acknowledgements The author is truly grateful to Andrea Braides and Johannes Zimmer for introducing her to the field of Variational Evolutions Problems and for sharing ideas, experiences and different point of views. This work was partially supported by the "National Group for Mathematical Analysis, Probability and their Applications" (GNAMPA-INDAM).

References

1. Abeyaratne, R., Chu, C., James, R.D.: Kinetics of materials with wiggly energies: theory and application to the evolution of twinning microstructures in a Cu-Al-Ni shape memory alloy. Philos. Mag. A: Phys. Condens. Matter, Struct. Defects Mech. Prop. **73**, 457–497 (1996)
2. Ambrosio, L., Gigli, N., Savaré, G.: Gradient Flows in Metric Spaces and in the Space of Probability Measures. Lectures in Mathematics ETH Zürich, 2nd edn. Birkhäuser Verlag, Basel (2008)
3. Ansini, N., Braides, A., Zimmer, J.: Minimising movements for oscillating energies: the critical regime. In: Proceedings of the Royal Society of Edinburgh A (to appear)
4. Braides, A.: Local Minimization, Variational Evolution and Γ-convergence. Lecture Notes in Mathematics, vol. 2094. Springer, Cham (2014)
5. Braides, A., Truskinovsky, L.: Asymptotic expansions by Γ-convergence. Continuum Mech. Thermodyn. **20**, 21–62 (2008)
6. Braides, A., Colombo, M., Gobbino, M., Solci, M.: Minimizing movements along a sequence of functionals and curves of maximal slope. C. R. Acad. Sci. Paris, Ser. I **354**, 685–689 (2016)
7. De Giorgi, E.: New problems on minimizing movements. In: Boundary Value Problems for Partial Differential Equations and Applications. RMA Res. Notes Appl. Math., vol. 29, pp. 81–98. Masson, Paris (1993)

Energy-Stable Numerical Schemes for Multiscale Simulations of Polymer–Solvent Mixtures

Mária Lukáčová-Medviďová, Burkhard Dünweg, Paul Strasser and Nikita Tretyakov

Abstract We present a new second-order energy dissipative numerical scheme to treat macroscopic equations aiming at the modeling of the dynamics of complex polymer–solvent mixtures. These partial differential equations are the Cahn-Hilliard equation for diffuse interface phase fields and the Oldroyd-B equations for the hydrodynamics of the polymeric mixture. A second-order combined finite volume/finite difference method is applied for the spatial discretization. A complementary approach to study the same physical system is realized by simulations of a microscopic model based on a hybrid Lattice Boltzmann/Molecular Dynamics scheme. These latter simulations provide initial conditions for the numerical solution of the macroscopic equations. This procedure is intended as a first step toward the development of a multiscale method that aims at combining the two models.

1 Introduction

Phase separation in binary fluids is a fundamental process in condensed matter physics. For Newtonian fluids, the phenomenon of spinodal decomposition is reasonably well understood in terms of the so-called "model H" [3, 10, 15], where

The present paper has been supported by the German Science Foundation (DFG) under the grant TRR 146.

M. Lukáčová-Medviďová (✉) · P. Strasser
Institute of Mathematics, University of Mainz, Mainz, Germany
e-mail: lukacova@uni-mainz.de

P. Strasser
e-mail: strasser@uni-mainz.de

B. Dünweg · N. Tretyakov
Max Planck Institute for Polymer Research, Mainz, Germany
e-mail: duenweg@mpip-mainz.mpg.de

N. Tretyakov
e-mail: tretyakov@mpip-mainz.mpg.de

P. van Meurs et al. (eds.), *Mathematical Analysis of Continuum Mechanics and Industrial Applications II*, Mathematics for Industry 30,
https://doi.org/10.1007/978-981-10-6283-4_13

the hydrodynamic equations of motion for mass and momentum conservation are coupled to a convection-diffusion equation for the concentration (or in general the "phase field" variable ϕ), and the thermodynamics, which is described by a (free) energy functional $E(\phi)$, gives rise to a driving force, see, e.g., [1, 9, 19, 21]. In such "diffuse interface" or "phase field" models, the interface between two phases is a thin layer of finite thickness, across which ϕ varies continuously. A big advantage of such models is that interfaces are defined implicitly and do not need to be tracked. Similarly, topological changes of the interface structure are automatically described correctly. However, phase field models are generally very challenging to solve numerically.

The physics (and therefore also the mathematics and numerics) becomes much more involved if one component—or both—is a macromolecular compound. In this case, the large molecular relaxation time gives rise to a dynamic coupling between intra-molecular processes and the unmixing on experimentally relevant time scales, with interesting new phenomena, for which the term "viscoelastic phase separation" [18] has been coined. Here the construction of physically sound dynamic equations with suitable constitutive relations to describe the viscoelasticity is already a challenge in itself. Tanaka [18] made the first attempt in this direction; however, Zhou et al. [21] showed later that this dynamics violates the second law of thermodynamics and provided a corrected set of equations that satisfy it. Nevertheless, a fully satisfactory solution of the problem is probably still missing. For this reason, we wish to carefully investigate how well this system describes the physics, by comparing it with and linking it to a computer experiment that is based upon a microscopic (molecular) model (see Sect. 4) that can be considered as physically sound beyond reasonable doubt. We thus study the diffuse-interface viscoelastic equations put forward in [21] for the case of the unmixing process of a polymer–solvent system.

Typically the interfacial region separating the two fluids is very narrow, and a high spatial resolution is required to accurately capture the interface dynamics. In fact, the underlying problem is stiff, which necessitates an implicit time discretization. Moreover, the solution admits several time scales over which it evolves, cf. [12]. In the literature, one can find already several numerical methods that have been used for the numerical approximation of diffuse interface models, see, e.g., [4, 9, 12, 13, 19] and the references therein.

In order to describe the dynamics of a complex polymer–solvent mixture, the Cahn–Hilliard equations for the phase field evolution are coupled with the Oldroyd-B equations, which consist of the momentum equation for the velocity field, the continuity equation, and the rheological equation for time evolution of the elastic stress tensor. We note in passing that there is quite a large number of analytical as well as numerical results available in the literature for the Oldroyd-B system, see, e.g., [2, 6, 7, 14]. The main challenge in this field is to obtain a stable approximate numerical solution for large Weissenberg numbers. The dimensionless Weissenberg number represents elastic effects; it is large when the molecular relaxation time is comparable to the time scale of the flow, or even exceeds it significantly. In the present work, we consider the non-critical regime of Weissenberg numbers. Applying the techniques from [6, 14], a further generalization using the log-transformation of the elastic stress tensor and the Lagrange-type approximation of the convective term is possible.

2 Mathematical Model

Following Zhou et al. [21], the total energy of the polymer–solvent mixture consists of the mixing, the polymerization, the elastic, and the kinetic energy:

$$\begin{aligned} E_{tot}(\phi, q, \sigma, \mathbf{u}) &= E_{mix}(\phi) + E_{pol}(q) + E_{el}(\sigma) + E_{kin}(\mathbf{u}) \\ &= \int_\Omega \left(\frac{C_0}{2}|\nabla\phi|^2 + F(\phi)\right) + \int_\Omega \frac{1}{2}|q|^2 + \int_\Omega \frac{1}{2}\mathrm{tr}(\sigma) + \int_\Omega \frac{1}{2}|\mathbf{u}|^2. \end{aligned} \tag{1}$$

Here ϕ is the volume fraction of polymer molecules, q the bulk stress, σ the elastic shear stress tensor, and $\mathbf{u}$ the volume averaged velocity. Furthermore, C_0 is a positive constant and $F(\phi)$ denotes a double-well potential. In the present paper, we work with the logarithmic potential derived from standard Flory–Huggins theory [16]

$$F(\phi) = \frac{1}{n_p}\phi\ln\phi + \frac{1}{n_s}(1-\phi)\ln(1-\phi) + \chi\phi(1-\phi) \tag{2}$$

with n_p and n_s the molecular weight of the two components (p: polymer, s: solvent) and $\chi = \chi_0/T$ the effective Flory interaction which depends on the temperature T. Consequently, $\phi \in [0, 1]$.

Following standard procedures in non-equilibrium thermodynamics, Zhou et al. [21] split the currents into a reversible and a nonreversible part. The reversible contributions are obtained through the virtual work principle, while the irreversible contributions are obtained by analyzing the dissipative process. The corresponding dynamic model then reads

$$\begin{aligned} \frac{\partial\phi}{\partial t} + \mathbf{u}\cdot\nabla\phi &= \nabla\cdot\left\{\phi(1-\phi)\,M\Big[\phi(1-\phi)\nabla\mu - \nabla(A_1(\phi)\,q)\Big]\right\}, \\ \frac{\partial q}{\partial t} + \mathbf{u}\cdot\nabla q &= -\frac{1}{\tau(\phi)}q - A_1(\phi)\,\nabla\cdot\left\{M\Big[\phi(1-\phi)\nabla\mu - \nabla(A_1(\phi)\,q)\Big]\right\}, \\ \frac{\partial\sigma}{\partial t} + (\mathbf{u}\cdot\nabla)\sigma &= (\nabla\mathbf{u})\cdot\sigma + \sigma\cdot(\nabla\mathbf{u})^T - \frac{1}{\tau_s(\phi)}\sigma + B_2(\phi)\Big[\nabla\mathbf{u} + (\nabla\mathbf{u})^T\Big], \\ \frac{\partial\mathbf{u}}{\partial t} + (\mathbf{u}\cdot\nabla)\mathbf{u} &= -\nabla p + \nabla\cdot\left\{\eta(\phi)\Big[\nabla\mathbf{u} + (\nabla\mathbf{u})^T\Big]\right\} + \mu\nabla\phi + \nabla\cdot\sigma, \\ \nabla\cdot\mathbf{u} &= 0, \end{aligned} \tag{3}$$

where $M > 0$ is the mobility coefficient, $\tau(\phi) = \tau^0\phi^2$ and $\tau_s(\phi) = \tau_s^0\phi^2$ are the relaxation times, $B_2(\phi) = m_s^0\phi^2$ is the relaxation modulus, $A_1(\phi) \in [1, 2]$ is the bulk modulus, $\eta(\phi) \in [0, 1]$ is the viscosity and all relaxation coefficients are positive. Furthermore, p denotes the pressure and μ, the chemical potential, is given by

$$\mu = \frac{\delta E_{mix}(\phi)}{\delta \phi} = -C_0 \Delta \phi + F'(\phi)\,. \tag{4}$$

This model is thermodynamically consistent and dissipates energy over time if $\mathrm{tr}(\sigma) > 0$. The latter holds true at any time, if the determinant of the initial value of the so-called conformation tensor $\mathbf{c}$ is greater than one, see Hu and Lelièvre [11]. Here $\mathbf{c}$ is defined as $\mathbf{c} := \frac{1}{B_2(\phi)}\sigma + \mathbb{1}$, where $\mathbb{1}$ is the identity matrix.

Theorem 1 *The problem* (3) *satisfies the following energy law*

$$\begin{aligned}
&\frac{dE_{tot}(\phi, q, \sigma, \mathbf{u})}{dt} + \frac{1}{\tau^0}\left\|\frac{q}{\phi}\right\|^2_{L^2(\Omega)} + \int_\Omega M\Big[\phi(1-\phi)\nabla\mu - \nabla(A_1(\phi)\,q)\Big]^2 \\
&+ \int_\Omega \frac{1}{2\,\tau_s(\phi)}\mathrm{tr}(\sigma) + \int_\Omega \frac{\eta(\phi)}{2}\sum_{i,j}\left(\frac{\partial u_i}{\partial x_j} + \frac{\partial u_j}{\partial x_i}\right)^2 = 0\,.
\end{aligned} \tag{5}$$

Proof Multiplying $(3)_1$ by μ and integrating over the computational domain Ω, assuming suitable boundary conditions (e.g., periodic boundary conditions), and applying integration by parts, we obtain

$$\begin{aligned}
&\int_\Omega \frac{\partial\phi}{\partial t}\mu + \int_\Omega \mathbf{u}\cdot\nabla\phi\mu - \int_\Omega \nabla\cdot\Big\{\phi(1-\phi)M\Big[\phi(1-\phi)\nabla\mu - \nabla(A_1(\phi)\,q)\Big]\Big\}\,\mu \\
&= \int_\Omega \frac{\partial\phi}{\partial t}\frac{\delta E_{mix}(\phi)}{\delta\phi} + \int_\Omega \mathbf{u}\cdot\nabla\phi\mu + \int_\Omega \Big\{\phi(1-\phi)M\Big[\phi(1-\phi)\nabla\mu - \nabla(A_1(\phi)\,q)\Big]\Big\}\,\nabla\mu \\
&= \frac{dE_{mix}(\phi)}{dt} + \int_\Omega \mathbf{u}\cdot\nabla\phi\mu + \int_\Omega M\Big[\phi(1-\phi)\nabla\mu - \nabla(A_1(\phi)\,q)\Big]\Big[\phi(1-\phi)\nabla\mu\Big] = 0\,.
\end{aligned}$$

Now, using the previously explained procedure after multiplying $(3)_2$ by q, and making use of $\nabla\cdot\mathbf{u} = 0$ yield

$$\begin{aligned}
&\int_\Omega \frac{\partial q}{\partial t}q + \int_\Omega \mathbf{u}\cdot\nabla q q + \int_\Omega \frac{1}{\tau(\phi)}q^2 + \int_\Omega A_1(\phi)\,\nabla\cdot\Big\{M\Big[\phi(1-\phi)\nabla\mu - \nabla(A_1(\phi)\,q)\Big]\Big\}\,q \\
&= \int_\Omega \frac{1}{2}\frac{\partial}{\partial t}q^2 + \frac{1}{2}\int_\Omega \mathbf{u}\cdot\nabla q^2 + \int_\Omega \frac{1}{\tau^0\phi^2}q^2 \\
&\qquad + \int_\Omega \nabla\cdot\Big\{M\Big[\phi(1-\phi)\nabla\mu - \nabla(A_1(\phi)\,q)\Big]\Big\}\,A_1(\phi)\,q \\
&= \frac{d}{dt}\left(\int_\Omega \frac{1}{2}q^2\right) - \frac{1}{2}\int_\Omega \nabla\cdot\mathbf{u}\,q^2 + \frac{1}{\tau^0}\left\|\frac{q}{\phi}\right\|^2_{L^2(\Omega)} \\
&\qquad - \int_\Omega M\Big[\phi(1-\phi)\nabla\mu - \nabla(A_1(\phi)\,q)\Big]\nabla(A_1(\phi)\,q) \\
&= \frac{dE_{pol}(q)}{dt} + \frac{1}{\tau^0}\left\|\frac{q}{\phi}\right\|^2_{L^2(\Omega)} + \int_\Omega M\Big[\phi(1-\phi)\nabla\mu - \nabla(A_1(\phi)\,q)\Big]\Big[-\nabla(A_1(\phi)\,q)\Big] = 0\,.
\end{aligned}$$

For the next step we remind that for all $A \in \mathbb{R}^{n\times n},\quad n \in \mathbb{N}:\quad \mathrm{tr}(\mathbf{A}) = \mathrm{tr}(\mathbf{A}\cdot \mathbb{1}^T) = \mathbf{A} : \mathbb{1}$. Hence, applying the aforementioned computational steps after multiplying $(3)_3$ by $\frac{1}{2}\mathbb{1}$ implies

$$\begin{aligned}
&\int_\Omega \frac{1}{2}\mathrm{tr}\left(\frac{\partial\sigma}{\partial t}\right) + \int_\Omega \frac{1}{2}\mathrm{tr}\left((\mathbf{u}\cdot\nabla)\sigma\right) - \int_\Omega \frac{1}{2}\left(\nabla\mathbf{u} : \sigma^T + \sigma : \nabla\mathbf{u}\right) + \int_\Omega \frac{1}{2\,\tau_s(\phi)}\mathrm{tr}(\sigma) \\
&\qquad\qquad - \int_\Omega B_2(\phi)\mathrm{tr}(\nabla\mathbf{u}) \\
&= \int_\Omega \frac{1}{2}\frac{\partial}{\partial t}\mathrm{tr}(\sigma) - \int_\Omega \frac{1}{2}\mathrm{tr}\left((\nabla\cdot\mathbf{u})\sigma\right) - \int_\Omega \sigma : \nabla\mathbf{u} + \int_\Omega \frac{1}{2\,\tau_s(\phi)}\mathrm{tr}(\sigma) - \int_\Omega B_2(\phi)(\nabla\cdot\mathbf{u}) \\
&= \frac{d}{dt}\int_\Omega \frac{1}{2}\mathrm{tr}(\sigma) - \int_\Omega \sigma : \nabla\mathbf{u} + \int_\Omega \frac{1}{2\,\tau_s(\phi)}\mathrm{tr}(\sigma) = 0\,.
\end{aligned}$$

Multiplying $(3)_4$ by $\mathbf{u}$ (scalar product) and again applying the same procedure give

$$\begin{aligned}
&\int_\Omega \frac{\partial\mathbf{u}}{\partial t}\cdot\mathbf{u} + \frac{1}{2}\int_\Omega (\mathbf{u}\cdot\nabla)|\mathbf{u}|^2 + \int_\Omega \left\{\eta(\phi)\left[\nabla\mathbf{u} + (\nabla\mathbf{u})^T\right]\right\} : \nabla\mathbf{u} - \int_\Omega p(\nabla\cdot\mathbf{u}) \\
&\qquad\qquad - \int_\Omega \mu\nabla\phi\cdot\mathbf{u} - \int_\Omega \nabla\cdot\sigma\cdot\mathbf{u} \\
&= \int_\Omega \frac{1}{2}\frac{\partial|\mathbf{u}|^2}{\partial t} - \frac{1}{2}\int_\Omega (\nabla\cdot\mathbf{u})|\mathbf{u}|^2 + \int_\Omega \left\{\eta(\phi)\left[|\nabla\mathbf{u}|^2 + \mathrm{tr}\left((\nabla\mathbf{u})^2\right)\right]\right\} \\
&\qquad\qquad - \int_\Omega \mathbf{u}\cdot\nabla\phi\mu + \int_\Omega \sigma : \nabla\mathbf{u} \\
&= \frac{d}{dt}\int_\Omega \frac{1}{2}|\mathbf{u}|^2 + \int_\Omega \frac{\eta(\phi)}{2}\sum_{i,j}\left(\frac{\partial u_i}{\partial x_j} + \frac{\partial u_j}{\partial x_i}\right)^2 - \int_\Omega \mathbf{u}\cdot\nabla\phi\mu + \int_\Omega \sigma : \nabla\mathbf{u} = 0\,.
\end{aligned}$$

Summing up all four equations yields the energy law (5). We refer a reader also to [21], where analogous energy dissipation property has stated. □

3 Numerical Scheme

In this section, we present a new combined finite volume-finite difference method for the numerical solution of the macroscopic equations (3). We consider a uniform partition of the time interval $[0, T]$ with constant time step size Δt. In the first step, we propose to apply a linear second-order time discretizaton in order to approximate the crucial phase field variables (ϕ, q) of the Cahn–Hilliard-type equations.

Step 1. Find (ϕ^{n+1}, q^{n+1}), such that

$$
\begin{aligned}
&\frac{\phi^{n+1}-\phi^{n}}{\Delta t}+\tilde{\mathbf{u}}^{n}\cdot\nabla\tilde{\phi}^{n}=\nabla\cdot\left\{\tilde{\phi}^{n}(1-\tilde{\phi}^{n})M\Big[\tilde{\phi}^{n}(1-\tilde{\phi}^{n})\nabla\mu^{n+\frac{1}{2}}-\nabla(A_1(\tilde{\phi}^{n})\,q^{n+\frac{1}{2}})\Big]\right\},\\
&\mu^{n+\frac{1}{2}}=-C_0\Delta\phi^{n+\frac{1}{2}}+f(\phi^{n+1},\phi^{n}),\\
&\frac{q^{n+1}-q^{n}}{\Delta t}+\mathbf{u}^{n}\cdot\nabla q^{n+\frac{1}{2}}=\\
&=-\frac{1}{\tau(\tilde{\phi}^{n})}q^{n+\frac{1}{2}}-A_1(\tilde{\phi}^{n})\nabla\cdot\left\{M\Big[\tilde{\phi}^{n}(1-\tilde{\phi}^{n})\nabla\mu^{n+\frac{1}{2}}-\nabla(A_1(\tilde{\phi}^{n})\,q^{n+\frac{1}{2}})\Big]\right\},
\end{aligned}
\tag{6}
$$

where $\phi^{n+\frac{1}{2}}=\frac{1}{2}(\phi^{n+1}+\phi^{n})$ and $q^{n+\frac{1}{2}}=\frac{1}{2}(q^{n+1}+q^{n})$. Further,

$$
\tilde{\mathbf{u}}^{n}:=\mathbf{u}^{n}+\Delta t\,\mu^{n+\frac{1}{2}}\nabla\tilde{\phi}^{n}, \tag{7}
$$

to split the phase field part from the hydrodynamic part and $\tilde{\phi}^{n}=\frac{1}{2}(3\phi^{n}-\phi^{n-1})$ to keep the second-order time discretization while staying linear, see Tierra and Guillén-González [19]. For the approximation of the derivative of the double-well potential, we set

$$
f(\phi^{n+1},\phi^{n})=f(\phi^{n})+\frac{1}{2}f'(\phi^{n})\,(\phi^{n+1}-\phi^{n}).
$$

This discretization is the so-called optimal dissipation approximation which ensures a second-order approximation for smooth phase fields and entropy dissipation [19].

In the second step, we approximate the fluid equations as follows.

Step 2. Denote by $D(\mathbf{u}^{n+1})=\frac{1}{2}\Big[\nabla\mathbf{u}^{n+1}+(\nabla\mathbf{u}^{n+1})^{T}\Big]$ the symmetric velocity gradient. Applying the Chorin pressure projection method, we find $(\mathbf{u}^{n+1},p^{n+1})$, such that

$$
\begin{aligned}
&\frac{\mathbf{u}^{n+1}-\tilde{\mathbf{u}}^{n}}{\Delta t}+(\mathbf{u}^{n}\cdot\nabla)\mathbf{u}^{n+1}=\nabla\cdot\left\{\eta(\phi^{n+\frac{1}{2}})\,2D(\mathbf{u}^{n+1})\right\}-\nabla p^{n+1}+\nabla\cdot\sigma^{n},\\
&\nabla\cdot\mathbf{u}^{n+1}=0.
\end{aligned}
\tag{8}
$$

Note that (8) yields a first-order time implicit discretization. Generalization to higher orders is possible by applying, e.g., the fractional θ-scheme [20].

Finally, in the third step, we approximate the Oldroyd-B equation for the time evolution of the viscoelastic stress tensor σ.

Step 3. Find σ^{n+1}, such that

$$
\begin{aligned}
\frac{\sigma^{n+1}-\sigma^{n}}{\Delta t}+(\mathbf{u}^{n+1}\cdot\nabla)\sigma^{n}&=(\nabla\mathbf{u}^{n+1})\cdot\sigma^{n}-\sigma^{n}\cdot\left(\nabla\mathbf{u}^{n+1}\right)^{T}\\
&\quad-\frac{1}{\tau_s(\phi^{n+\frac{1}{2}})}\sigma^{n}+B_2(\phi^{n+\frac{1}{2}})\,2D(\mathbf{u}^{n+1}).
\end{aligned}
\tag{9}
$$

Theorem 2 *The numerical scheme* (6)–(9) *is energy stable. More precisely, the scheme satisfies the discrete version of the energy law*

$$\frac{E_{tot}(\phi^{n+1}, q^{n+1}, \sigma^{n+1}, \mathbf{u}^{n+1}) - E_{tot}(\phi^n, q^n, \sigma^n, \mathbf{u}^n)}{\Delta t} + \int_\Omega \frac{1}{2\,\tau_s(\phi^{n+\frac{1}{2}})} \mathrm{tr}(\sigma^n)$$
$$+ \int_\Omega \frac{\eta(\phi^{n+\frac{1}{2}})}{2} \sum_{i,j} \left(\frac{\partial u_i^{n+1}}{\partial x_j} + \frac{\partial u_j^{n+1}}{\partial x_i} \right)^2 + \frac{1}{\tau^0} \left\| \frac{q^{n+\frac{1}{2}}}{\tilde{\phi}^n} \right\|^2_{L^2(\Omega)}$$
$$+ \int_\Omega M \Big[\tilde{\phi}^n (1 - \tilde{\phi}^n) \nabla \mu^{n+\frac{1}{2}} - \nabla (A_1(\tilde{\phi}^n)\, q^{n+\frac{1}{2}}) \Big]^2 + ND^{n+1}_{phobic} + ND^{n+1}_{split} = 0\,, \tag{10}$$

where

$$ND^{n+1}_{phobic} := \int_\Omega f(\phi^{n+1}, \phi^n) \frac{\phi^{n+1} - \phi^n}{\Delta t} - \int_\Omega \frac{F(\phi^{n+1}) - F(\phi^n)}{\Delta t}\,,$$

$$ND^{n+1}_{split} := \frac{1}{2\,\Delta t} \Big(\|\mathbf{u}^{n+1} - \tilde{\mathbf{u}}^n\|^2_{L^2(\Omega)} + \|\tilde{\mathbf{u}}^n - \mathbf{u}^n\|^2_{L^2(\Omega)} \Big)\,.$$

Proof Similar to the proof of the continuous energy law, we multiply $(6)_1$ by $\mu^{n+\frac{1}{2}}$, $(6)_3$ by $q^{n+\frac{1}{2}}$, $(8)_1$ by $\mathbf{u}^{n+1}$, and (9) by $\frac{1}{2}\mathbb{1}$, and integrate. Assuming suitable boundary conditions, the calculation of the discrete polymeric and elastic energy is analogous to the continuous case. The calculation of the discrete mixing energy leads to the additional terms ND^{n+1}_{phobic} and $\int_\Omega (\tilde{\mathbf{u}}^n \cdot \nabla \tilde{\phi}^n) \mu^{n+\frac{1}{2}}$. The key point of the splitting scheme lies in the matching of the latter term with $\int_\Omega \frac{1}{\Delta t}(\mathbf{u}^{n+1} - \tilde{\mathbf{u}}^n) \cdot \mathbf{u}^{n+1}$, which arises while calculating the discrete kinetic energy. This is possible by multiplying expression (7) by $\tilde{\mathbf{u}}^n$ and integrating which yield

$$\|\tilde{\mathbf{u}}^n\|^2_{L^2(\Omega)} = \int_\Omega \mathbf{u}^n \cdot \tilde{\mathbf{u}}^n + \int_\Omega \Delta t (\tilde{\mathbf{u}}^n \cdot \nabla \tilde{\phi}^n) \mu^{n+\frac{1}{2}}\,.$$

This implies

$$\begin{aligned}\int_\Omega (\tilde{\mathbf{u}}^n \cdot \nabla \tilde{\phi}^n) \mu^{n+\frac{1}{2}} &= \frac{1}{\Delta t} \left(\|\tilde{\mathbf{u}}^n\|^2_{L^2(\Omega)} - \int_\Omega \mathbf{u}^n \cdot \tilde{\mathbf{u}}^n \right) \\ &= \frac{1}{2\,\Delta t} \Big(\|\tilde{\mathbf{u}}^n\|^2_{L^2(\Omega)} - \|\mathbf{u}^n\|^2_{L^2(\Omega)} + \|\tilde{\mathbf{u}}^n - \mathbf{u}^n\|^2_{L^2(\Omega)} \Big)\,,\end{aligned}$$

since

$$\int_\Omega \frac{1}{\Delta t} (\mathbf{u}^{n+1} - \tilde{\mathbf{u}}^n) \cdot \mathbf{u}^{n+1} = \frac{1}{2\,\Delta t} \Big(\|\mathbf{u}^{n+1}\|^2_{L^2(\Omega)} - \|\tilde{\mathbf{u}}^n\|^2_{L^2(\Omega)} + \|\mathbf{u}^{n+1} - \tilde{\mathbf{u}}^n\|^2_{L^2(\Omega)} \Big).$$

In conclusion, we obtain the discrete energy law (10) and consequently

$$E_{tot}(\phi^{n+1}, q^{n+1}, \sigma^{n+1}, \mathbf{u}^{n+1}) \le E_{tot}(\phi^n, q^n, \sigma^n, \mathbf{u}^n), \tag{11}$$

provided that we control ND^{n+1}_{phobic}, since all other terms are non-negative. In [9], it has been shown that $ND^{n+1}_{phobic} = \mathcal{O}(\Delta t^2)$. Our numerical experiments presented below indeed confirm that the energy dissipation (11) holds. □

Let us note that the computational domain $\Omega \subset \mathbb{R}^2$ is discretized by a regular rectangular grid. The discretization in space is realized by a finite volume scheme using second-order upwinding for the convective terms and central differences for the remaining terms.

4 Lattice Boltzmann and Molecular Dynamics Simulation

Starting point of the simulation model is a standard Kremer–Grest model [8], where polymer chains are represented by sequences of N_{ch} beads each. The beads interact via a bonded potential (FENE springs) in order to ensure connectivity, and a non-bonded potential to model the excluded volume effect as well as the quality of the solvent. In good solvent, the latter interaction is simply a purely repulsive Lennard-Jones (LJ) potential. For the form of these potentials, as well as the parameters, see [8]. The LJ potential also defines the unit system of the simulation (each bead has unit mass). The effects of less-than-perfect or even poor solvent quality are modeled by adding an attractive tail to the non-bonded pair interaction. For this tail, we take a suitably fitted cosine wave such that the range of the potential is 1.5 and its depth a parameter $\phi_{attr} \geq 0$, which is a direct measure of solvent quality. For details of that potential, see [17]. This system is simulated by Molecular Dynamics (MD) and at the same time coupled to a standard D3Q19 Lattice Boltzmann (LB) model. The latter represents the momentum transport through the solvent and has the thermodynamic properties of an ideal gas. The coupling is facilitated by a Stokes friction acting on each bead. The dissipative nature of the coupling ensures that it does not alter the thermodynamics of the polymer system. Both MD and LB are supplemented by a Langevin thermal noise such that the temperature is being kept constant. For technical details of that approach, as well as the underlying theory, see [5]. It should be noted that the simulations are done in three-dimensional space, while the solution of the macroscopic equations is done in 2D. The pure LB fluid is strictly Newtonian, and all non-Newtonian effects arise from the coupling to the polymer system.

5 Numerical Experiments

As a first numerical experiment, the dynamic equations were solved on our computational domain $\Omega = [0, 1] \times [0, 1]$, which is devided into 128×128 grid cells and has periodic boundaries. The initial data of the volume fraction $\phi(t = 0)$ is taken to be constant with a small random perturbation and the initial velocity and bulk stress are set to zero. The elastic stress tensor $\sigma(t = 0) = \sqrt{2}\,\mathbb{1}$, the mixing con-

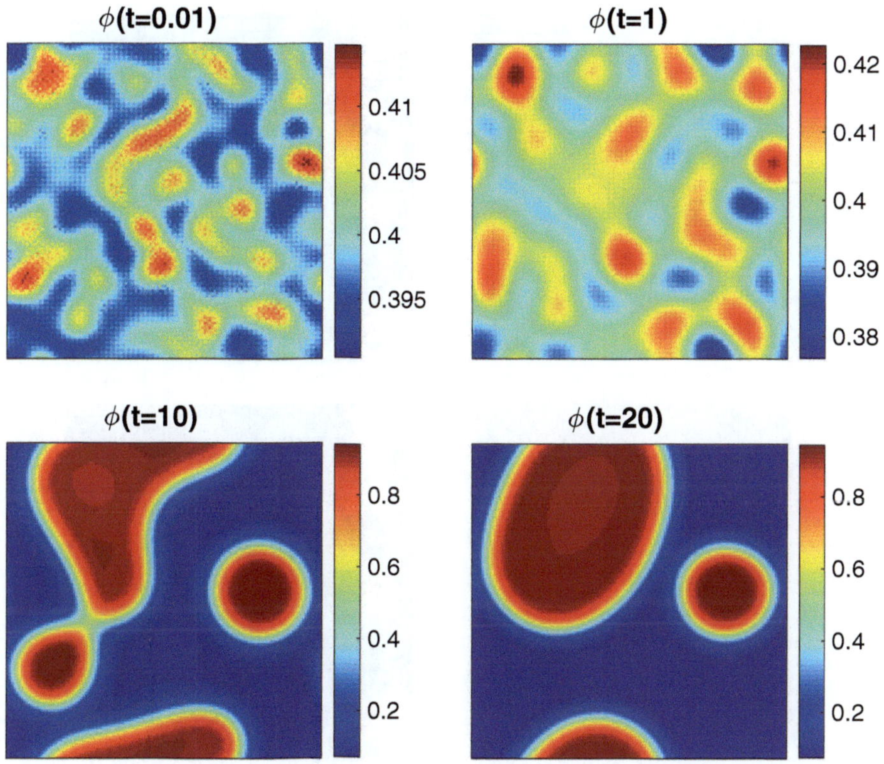

Fig. 1 128 × 128 grid with homogeneous initial volume fraction with noise: $\phi(t=0) = 0.4 + \delta, \quad \delta \in [-0.05, 0.05]$

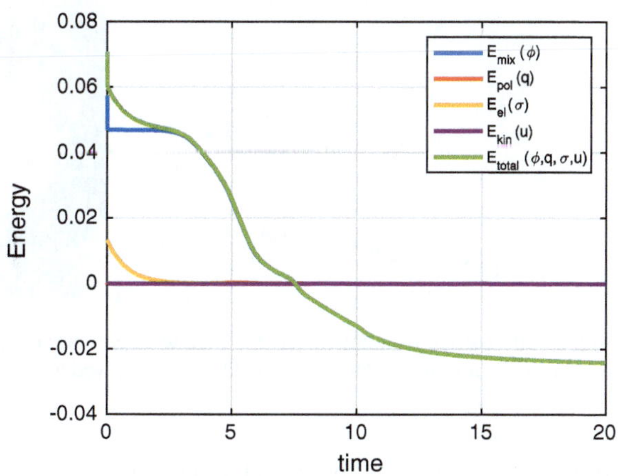

Fig. 2 Energy evolution in experiment 1

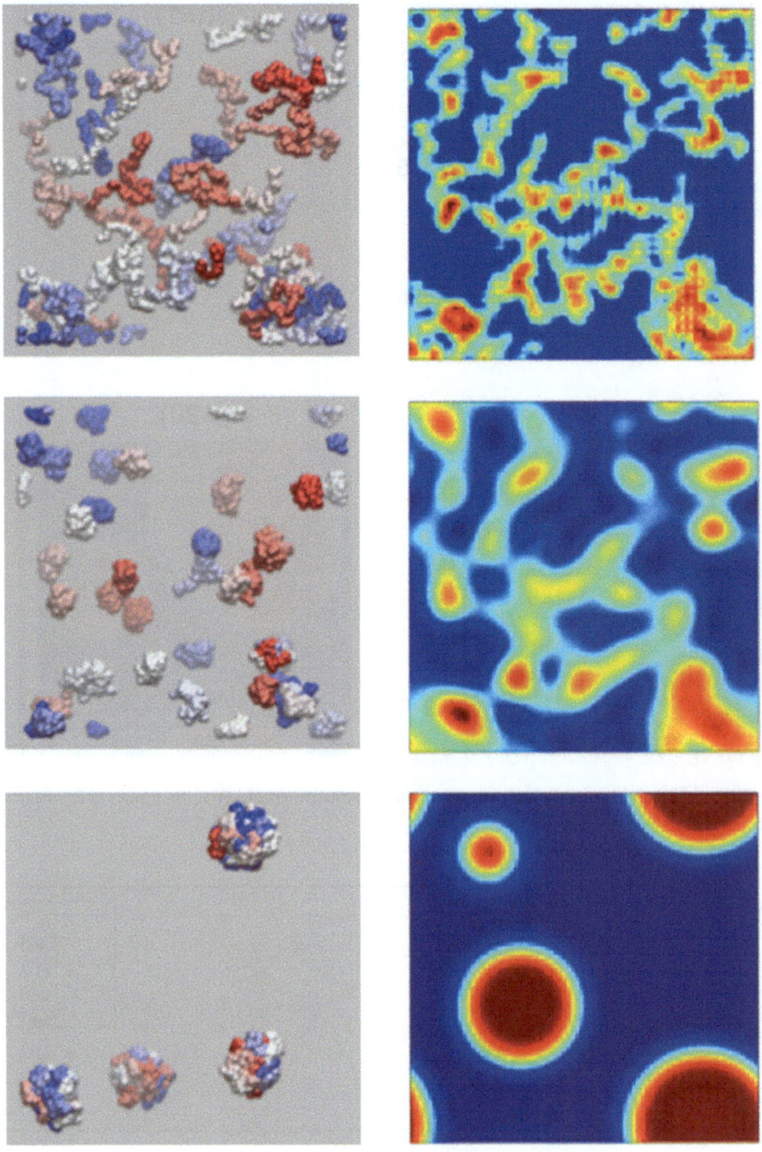

Fig. 3 80^3 Lattice Boltzmann solvent cube with polymer chains from Molecular Dynamics (first row, left column) mapped to 80×80 2D grid and solved with our numerical scheme (right column)

stant $C_0 = 1/600$, the molecular weights $n_p = n_s = 1$, the Flory constant $\chi_0 = 3.3$ and the temperature $T = 1.1$. Furthermore, the mobility coefficient $M = 10$, and the relaxation coefficients $\tau^0 = 10$, $\tau_s^0 = 5$ and $m_s^0 = 0.2$. The experiment as shown in Fig. 1 demonstrates phase separation by aggregation of polymer molecules toward droplets. The droplets merge over time causing noticeable decreases of the mixing energy, as shown in Fig. 2. Accordant to our proofs, the total energy is (strictly) monotonically decreasing over time, even while there is no merging, which is related to the surface minimization of the droplets.

As a second numerical experiment, we ran the LB/MD system for the following parameters: We study 40 chains of 100 beads each immersed in a volume of size 80^3, such that the volume fraction is roughly 4×10^{-3}, assuming a bead diameter of one LJ unit. The system is equilibrated in good solvent conditions ($\phi_{attr} = 0$) at temperature $T = 1$ (in units where Boltzmann's constant is one). Coupling to the LB system is facilitated by a Stokes friction $\zeta = 20$ and nearest-neighbor velocity interpolation. The MD time step is 5×10^{-3} and the LB time step ten times larger. The LB lattice spacing is one, and the LB shear viscosity is four. After equilibration, ϕ_{attr} is suddenly increased to one, which is in the poor solvent regime. Figure 3, left column, shows the time development of the system after this "quench."

The starting configuration of the 3D simulation was then transferred to the 2D solver of the macroscopic equations by subdividing the 3D system into columns of size $1 \times 1 \times 80$, counting the number of beads in each column and averaging their velocities weighted with the solvent velocity field. By such a projection, we obtain an effective 2D volume fraction, which is, on average, roughly 0.3 and an effective 2D velocity field. Further, parameters of the macroscopic solver are unchanged.

The numerical scheme (right column in Fig. 3) behaves analogous to the first experiment. Starting with initial configuration (at the top), the polymer concentration is then being smoothed (middle) and the system starts to phase-separate by aggregation into larger droplets (bottom). The 3D MD/LB simulations (left column in Fig. 3) demonstrate a similar process with another level of details. The polymer chains in the initial configuration at good solvent condition form coils (top). After quenching the system into poor solvent regime, the individual chains collapse (middle) and later form larger droplets by aggregation (bottom).

Although the mapping is of course extremely crude, it is nevertheless clear that the two systems evolve at least somewhat similarly. The relations for mass and energy scales between two models can be established by matching volume fractions and Flory parameters. However, a detailed comparison of time scales has not yet been accomplished: Though the viscosity of the fluid in the numerical model is known, at least a tiny fraction of the polymer is always present in every grid cell. Only averaging over these cells will give us the desired value of the effective viscosity needed to match the time scales. This analysis is left for future work.

6 Outlook

The present work is intended only as a first step toward much more detailed investigations. It is hoped on the one hand that in the future the underlying physics of the dynamic equations will become more clear, and on the other hand that much more detailed and stringent comparisons will provide deeper insight and perhaps also the development of more refined and accurate macroscopic models, with the possible perspective of even constructing multiscale models, which would consist of hybrid schemes that incorporate aspects of both the approaches presented here.

References

1. Abels, H., Depner, D., Garcke, H.: Existence of weak solutions for a diffuse interface model for two-phase flows of incompressible fluids with different densities. J. Math. Fluid Mech. **15**(3), 453–480 (2013). https://doi.org/10.1007/s00021-012-0118-x
2. Barrett, J.W., Boyaval, S.: Existence and approximation of a (regularized) Oldroyd-B model. Math. Models Methods Appl. Sci. **21**(09), 1783–1837 (2011). https://doi.org/10.1142/S0218202511005581
3. Bray, A.J.: Theory of phase-ordering kinetics. Adv. Phys. **51**(2), 481–587 (2002). https://doi.org/10.1080/00018730110117433
4. Cheng, Y., Kurganov, A., Qu, Z., Tang, T.: Fast and stable explicit operator splitting methods for phase-field models. J. Comput. Phys. **303**, 45–65 (2015). https://doi.org/10.1016/j.jcp.2015.09.005
5. Dünweg, B., Ladd, A.J.C.: Lattice Boltzmann simulations of Soft Matter systems. In: Holm, C., Kremer, K. (eds.) Advanced Computer Simulation Approaches for Soft Matter Sciences III. Advances in Polymer Science, no. 221, pp. 89–166. Springer, Berlin (2009)
6. Fattal, R., Kupferman, R.: Time-dependent simulation of viscoelastic flows at high Weissenberg number using the log-conformation representation. J. Non-Newton. Fluid Mech. **126**(1), 23–37 (2005). https://doi.org/10.1016/j.jnnfm.2004.12.003
7. Fernández-Cara, E., Guillén-González, F.M., Ortega, R.R.: Mathematical modeling and analysis of viscoelastic fluids of the Oldroyd kind. In: Handbook of Numerical Analysis, vol. VIII, pp. 543–660. Elsevier (2002)
8. Grest, G.S., Kremer, K.: Molecular dynamics simulation for polymers in the presence of a heat bath. Phys. Rev. A **33**(5), 3628–3631 (1986). https://doi.org/10.1103/PhysRevA.33.3628
9. Guillén-González, F., Tierra, G.: On linear schemes for a Cahn-Hilliard diffuse interface model. J. Comput. Phys. **234**, 140–171 (2013). https://doi.org/10.1016/j.jcp.2012.09.020
10. Hohenberg, P.C., Halperin, B.I.: Theory of dynamic critical phenomena. Rev. Mod. Phys. **49**(3), 435–479 (1977). https://doi.org/10.1103/RevModPhys.49.435
11. Hu, D., Lelièvre, T.: New entropy estimates for the Oldroyd-B model and related models. Commun. Math. Sci. **5**(4), 909–916 (2007)
12. Kay, D., Welford, R.: Efficient numerical solution of Cahn-Hilliard-Navier-Stokes fluids in 2d. SIAM J. Sci. Comput. **29**(6), 2241–2257 (2007). https://doi.org/10.1137/050648110
13. Lee, D., Huh, J.Y., Jeong, D., Shin, J., Yun, A., Kim, J.: Physical, mathematical, and numerical derivations of the Cahn-Hilliard equation. Comput. Mater. Sci. **81**, 216–225 (2014). https://doi.org/10.1016/j.commatsci.2013.08.027
14. Lukáčová-Medvid'ová, M., Notsu, H., She, B.: Energy dissipative characteristic schemes for the diffusive Oldroyd-B viscoelastic fluid. Int. J. Numer. Methods Fluids **81**(9), 523–557 (2016). https://doi.org/10.1002/fld.4195
15. Onuki, A.: Phase Transition Dynamics. Cambridge University Press, Cambridge (2002)
16. Rubinstein, M., Colby, R.H.: Polymer Physics. Oxford University Press, Oxford (2003)
17. Soddemann, T., Dünweg, B., Kremer, K.: A generic computer model for amphiphilic systems. Eur. Phys. J. E **6**(1), 409–419 (2001). https://doi.org/10.1007/s10189-001-8054-4
18. Tanaka, H.: Viscoelastic phase separation. J. Phys. Condens. Matter **12**(15), R207 (2000). https://doi.org/10.1088/0953-8984/12/15/201
19. Tierra, G., Guillén-González, F.: Numerical methods for solving the Cahn-Hilliard equation and its applicability to related energy-based models. Arch. Comput. Methods Eng. **22**(2), 269–289 (2015). https://doi.org/10.1007/s11831-014-9112-1

20. Turek, S., Ouazzi, A., Hron, J.: On pressure separation algorithms (PSepA) for improving the accuracy of incompressible flow simulations. Int. J. Numer. Methods Fluids **59**(4), 387–403 (2009). https://doi.org/10.1002/fld.1820
21. Zhou, D., Zhang, P., Weinan, E.: Modified models of polymer phase separation. Phys. Rev. E **73**(6), 061801 (2006). https://doi.org/10.1103/PhysRevE.73.061801

Part VI
Complex Systems

On Mathematical Modeling and Analysis of Brain Network

Hirotada Honda

Abstract In this article, we first formulate a functional equation-based modeling of the resting-state network, which is attracting attention in the research of the brain. Then, we discuss the local-in-time solvability of the model in a suitable function space.

1 Introduction

Research on the brain network has been done intensively in recent decades. Non-invasive measurement, such as fMRI, PET, and NIRS, enables us to measure and estimate the temporal neuronal activity in the brain of human beings.

At the same time, many mathematical models have been developed, mainly by using ordinary or partial differential equations. Although some of them have been intensively discussed, there are still a number of mathematical models and problems left unsolved.

In this article, we propose a temporal and macroscopic model of the resting-state network, which is closely related to the theory of weakly coupled oscillators. We present the well-posedness of the proposed model. Due to space limitations, the details of proofs will be shown in another full paper.

2 Resting-State Network

Recently, it was reported that synchronous neuronal firing between some specific regions in the brain emerges even when individuals are resting, and this was first named as the *default mode network* by Raichle [10, 11]. Later, similar phenomena

H. Honda (✉)
NTT Network Technology Laboratories, 3-9-11 Midori-Cho, Musashino-Shi,
Tokyo 180-8585, Japan
e-mail: honda.hirotada@lab.ntt.co.jp

P. van Meurs et al. (eds.), *Mathematical Analysis of Continuum Mechanics and Industrial Applications II*, Mathematics for Industry 30,
https://doi.org/10.1007/978-981-10-6283-4_14

were found in other regions, such as the sensory/motor component, executive control component, auditory component, and the temporal/parietal component [9]. We now use the term *resting-state network* to refer to the whole sets of synchronous neuronal firing during rest.

On the basis of these reports, Deco et al. [5] modeled the synchronous firing of neurons in the brain by using the Wilson–Cowan oscillators [15]. Later, Cabral et al. [1, 2] modeled it by using the Kuramoto model, which is the theory of weakly coupled oscillators in statistical physics.

Their proposal was the following system of ordinary equations,

$$\frac{\mathrm{d}}{\mathrm{d}t}\theta_m(t) = \omega_m + k\sum_{p=1}^{M} c_{mp}\sin\bigl(\theta_p(t-\tau_{mp}) - \theta_m(t)\bigr), \quad m = 1, 2, \ldots, M, \tag{1}$$

coupled with the definition of the order parameter

$$r(t)\mathrm{e}^{\mathrm{i}\Phi(t)} = \frac{1}{M}\sum_{m=1}^{M} \mathrm{e}^{\mathrm{i}\theta_m(t)}. \tag{2}$$

Hereafter, we use the notation of an imaginary unit, $\mathrm{i} = \sqrt{-1}$. In the above equations, the unknown variable is the average phase $\theta_n(t)$ of the n-th region, k, the unified coupling strength between each pair of regions; c_{mp}, the coupling strength between the m-th and p-th regions, and τ_{mp}, the transmission delay between the m-th and p-th regions

Although this model works well under the suitable choice of parameters, it is very time consuming to search for suitable values of parameters in a brute-forth manner. Therefore, we here try to apply a more macroscopic approach.

As for the statistical physics arguments, Ichinomiya [6] established a model that describes the behavior of weakly coupled oscillators on a random graph. For weakly coupled oscillators with delay, Lee [8] considered the infinite population limit of the Kuramoto model with delay.

However, we consider that the delay in the brain network depends on each axon, and so it is not purely randomized. On the basis of these considerations, we propose here the following functional equation, which corresponds to the Fokker–Planck equation of (1)–(2) with additive white noise. Hereafter, we use the notation $\Omega \equiv (0, 2\pi)$.

$$\begin{cases} \dfrac{\partial \rho}{\partial t} - D\dfrac{\partial^2 \rho}{\partial \phi^2} \\ \quad + \dfrac{\partial}{\partial \phi}\left[\rho(\phi, t; k, x, \omega)\left\{\omega + \dfrac{k}{\overline{P}}\displaystyle\int_{\mathbf{R}} G(x-y)\mathrm{d}y \int_{\mathbf{R}} g(\omega')\mathrm{d}\omega' \int_{\mathbf{R}_+} k'P(k')\mathrm{d}k' \right.\right. \\ \qquad \left.\left. \times \displaystyle\int_{\Omega} \rho(\phi', t-\tau(x,y); k', y, \omega')\Gamma(\phi-\phi')\,\mathrm{d}\phi'\right\}\right] = 0, \\ \qquad\qquad t > 0,\ (\phi, k, x, \omega) \in \Omega \times \mathbf{R}_+ \times \mathbf{R}^2, \\ \left.\dfrac{\partial^j \rho}{\partial \phi^j}\right|_{\phi=0} = \left.\dfrac{\partial^j \rho}{\partial \phi^j}\right|_{\phi=2\pi} \quad (j=0,1),\ t>0, \quad (k, x, \omega) \in \mathbf{R}_+ \times \mathbf{R}^2 \\ \rho\big|_{t=0} = \rho_0 \quad (\phi, k, x, \omega) \in \Omega \times \mathbf{R}_+ \times \mathbf{R}^2. \end{cases} \tag{3}$$

We also assume that $\rho = \rho_0$ for $t \leq 0$. Here, the independent variables are time t and phase ϕ, and the unknown variable is ρ, the probability density of phase. D is the diffusion coefficient of the white noise; $P(k)$ and $g(\omega)$, the probability distribution functions of the degree k and natural frequency ω, respectively; x and y, the location of nodes; $G(x-y)$, the coupling strength between nodes placed at x and y; $\tau(x, y)$, the delay between nodes placed at x and y, $\Gamma(\phi) = \sin(\phi)$, and $\overline{P} \equiv \int_{\mathbf{R}_+} kP(k)\,\mathrm{d}k$ is the expected value of the degree.

3 Function Spaces

In this subsection, we define the function spaces used throughout this paper. Let $T > 0$, and $\mathscr{G}$ be an open set in $\mathbf{R}$. Hereafter, $L_2(\mathscr{G})$ stands for a set of square-integrable functions defined on $\mathscr{G}$, equipped with the norm

$$\|u\| \equiv \int_{\mathscr{G}} |u(\phi)|^2\,\mathrm{d}\phi.$$

The inner product is defined by

$$(u_1, u_2) \equiv \int_{\mathscr{G}} u_1(\phi)\overline{u_2(\phi)}\,\mathrm{d}\phi,$$

where $\bar{z}$ denotes the complex conjugate of $z \in \mathbf{C}$.

For simplicity, we hereafter denote the L_2-norm of a function $f(\phi, t; k, x, \omega)$ with respect to ϕ merely by $\|f\|$ or $\|f(\cdot, t; k, x, \omega)\|$. Subject to the definition by Temam [13], we say, for a fixed set of parameters (k, x, ω), a 2π-periodic function

$$u(\phi; k, x, \omega) = \sum_{l=-\infty}^{\infty} a_l(k, x, \omega)e^{\mathrm{i}l\phi},$$

which is expanded in the Fourier series, belongs to the Sobolev space $\mathscr{H}^m(\mathscr{G})$ ($m \in \mathbf{R}_+$) when it satisfies

$$\|u(\cdot; k, x, \omega)\|_m^2 \equiv \sum_{l=-\infty}^{\infty} (1 + |l|^2)^m \left|a_l(k, x, \omega)\right|^2 < \infty.$$

Due to the definition of the Fourier series, the Fourier coefficients a_l ($l = \pm 1, \pm 2, \ldots$) of a function u are defined by

$$a_l(k, x, \omega) = \frac{1}{2\pi} \int_{\mathscr{G}} u(\phi; k, x, \omega) \mathrm{e}^{\mathrm{i}l\phi} \, \mathrm{d}\phi.$$

Note that, in the case of $m \in \mathbf{N}$, the norm above is equal to the usual Sobolev norm

$$\|u\|_{W_2^m(\mathscr{G})}^2 = \sum_{k \le m} \left\| \frac{\partial^k u}{\partial \phi^k} \right\|_{L_2(\mathscr{G})}^2.$$

For a Banach space E with the norm $\|\cdot\|_E$, we denote the space of E-valued measurable functions $u(t)$ on the interval (a, b) by $L_p(a, b; E)$, whose norm is defined by

$$\|u\|_{L_p(a,b;E)} \equiv \begin{cases} \left(\int_a^b \|u(t)\|_E^p \, \mathrm{d}t \right)^{1/p} & (1 \le p < \infty), \\ \operatorname*{ess\,sup}_{a \le t \le b} \|u(t)\|_E & p = \infty. \end{cases}$$

We also introduce the notation BUC to denote a set of bounded uniformly continuous functions with respect to its argument. Let us introduce the following notations:

$$\overline{\mathscr{H}^m} \equiv \left\{ u(\phi; k, x, \omega) \middle| |||u|||_m \equiv \sup_{k,x,\omega} \|u(\cdot; k, x, \omega)\|_m < \infty \right\},$$

$$L_1^{(1)} \equiv \left\{ u(\cdot; k, x, \omega) \in L_1(\Omega) \middle| u \ge 0, \int_\Omega u(\phi; k, x, \omega) \, \mathrm{d}\phi = 1 \ \forall (k, x, \omega) \right\},$$

$$L_1^{(1)}(T) \equiv \left\{ u(\cdot, t; k, x, \omega) \in L_1(\Omega) \middle| u \ge 0, \int_\Omega u(\phi, t; k, x, \omega) \, \mathrm{d}\phi = 1 \ \forall (t, k, x, \omega) \right\},$$

where $T > 0$ is an arbitrary number. In addition, we use the notations $|||u||| \equiv |||u|||_0$ for brevity. Hereafter, c's represent constants in the estimate of some quantities. When we denote $c(t)$ with suffixes, it is non-negative and depends on t non-decreasingly. We also define $C_l = \|\Gamma^{(l)}\|$ ($l = 0, 1, 2, \ldots$), where $f^{(l)}$ is the l-th times derivative of a function f in general.

For simplicity, we hereafter use the notations $u^{(j,k)} \equiv \left(\frac{\partial}{\partial \phi}\right)^j \left(\frac{\partial}{\partial t}\right)^k u$ ($j, k = 0, 1, 2, \ldots$) for a function $u = u(\phi, t)$ in general.

4 Main Results

In this section, we introduce our results concerning the solvability of problem (3). Our result is the local-in-time solvability of problem (3).

Theorem 1 *Let us assume* $\alpha > 1/2$ *and the following items:*

(i) $P(k) \in L_1(\mathbf{R}_+) \bigcap BUC(\mathbf{R}_+)$, $P(k) \geq 0$ *on* $k \in \mathbf{R}_+$, $\int_{\mathbf{R}_+} P(k)\,\mathrm{d}k = 1$ *and* $\int_{\mathbf{R}_+} kP(k)\mathrm{d}k < +\infty$*;*

(ii) $g(\omega) \in L_1(\mathbf{R}) \bigcap BUC(\mathbf{R})$, $g(\omega) \geq 0$ *on* $\omega \in \mathbf{R}$ *and* $\int_{\mathbf{R}} g(\omega)\,\mathrm{d}\omega = 1$*;*

(iii) $G(\cdot) \in BUC(\mathbf{R}) \bigcap L_1(\mathbf{R}) \bigcap L_{q'}(\mathbf{R})$, *where* $q' > 1$ *is specified later,* $G(x) \geq 0$ *and* $\int_{\mathbf{R}} G(x)\,\mathrm{d}x < M_1$ *with a positive constant* M_1*;*

(iv) $\rho_0 \in \overline{\mathscr{H}}^4 \bigcap L_1^{(1)}$*;*

(v) $\tau(x, y) \in BUC(\mathbf{R}^2)$, $\tau(x, y) \geq 0$ *on* $(x, y) \in \mathbf{R}^2$, *and* $0 < M_{21} < |\partial\tau/\partial y\,(x, y)| < M_{22}$ *with some constants* M_{2j} $(j = 1, 2)$.

Then, there exists a certain $T_* > 0$, $\alpha_j > 0$ $(j = 1, 2)$ *and a solution* $\rho(\phi, t; k, x, \omega) \in \mathscr{V}^4_{(\alpha_1,\alpha_2)}(T_*)$ *to (3) on* $(0, T_*)$, *where*

$$\mathscr{V}^4_{(\alpha_1,\alpha_2)}(T) \equiv \left\{\rho \in L_1^{(1)}(T) \middle| \mathrm{e}^{-\alpha_1 kt}\rho \in L_\infty(0, T; \overline{\mathscr{H}}^4) \bigcap C(0, T; \overline{\mathscr{H}}^2) \bigcap C^1(0, T; \overline{\mathscr{H}}^0) \right.$$

$$\left. \middle| \mathrm{e}^{-\alpha_2 kt}\rho^{(1,0)} \in L_2(0, T; \overline{\mathscr{H}}^4) \right\}.$$

Hereafter, for arbitrary $T > 0$, we use the notation

$$\mathscr{V}^4_{(\alpha_1,\alpha_2)(*)}(T) \equiv \left\{\rho \middle| \mathrm{e}^{-\alpha_1 kt}\rho \in L_\infty(0, T; \overline{\mathscr{H}}^4) \bigcap C(0, T; \overline{\mathscr{H}}^2) \bigcap C^1(0, T; \overline{\mathscr{H}}^0) \right.$$

$$\left. \middle| \mathrm{e}^{-\alpha_2 kt}\rho^{(1,0)} \in L_2(0, T; \overline{\mathscr{H}}^4) \right\}.$$

We carry out the proof of Theorem 1 in the three steps below:

(i) Existence of a solution ρ that belongs to $\mathscr{V}^4_{(\alpha_1,\alpha_2)(*)}(T_*)$ on a certain time interval $(0, T_*)$;

(ii) proof of $\rho \in L_1^{(1)}(T_*)$ and consequently of $\rho \in \mathscr{V}^4_{(\alpha_1,\alpha_2)}(T_*)$;

(iii) uniqueness of the solution in $\mathscr{V}^4_{(\alpha_1,\alpha_2)}(T_*)$.

4.1 Existence of Solution

We apply the semi-discrete approximation used by Sjöberg [12] and Tsutsumi [14] for the study of the KdV equation. Let us take $N \in \mathbf{N}$, $\delta = 2\pi/N$, $\phi_j = j\delta$ $(j = 1, 2, \ldots, N)$ and denote as ∂_+ and ∂_- the difference operators defined by

$$\delta\partial_+ f(\phi_j) = f(\phi_{j+1}) - f(\phi_j), \quad \delta\partial_- f(\phi_j) = f(\phi_j) - f(\phi_{j-1})$$

for a function f in general.

Now, instead of problem (3), we consider the following differential-difference equation:

$$\begin{cases} \dfrac{\partial \rho_N}{\partial t} - D\partial_+\partial_-\rho_N + \omega\partial_-\rho_N + k\partial_-(F_N\rho_N) = 0, \\ \rho_N(\phi_j, t) = \rho_N(\phi_{j+N}, t) \quad j = 1, 2, \ldots, N, \ t > 0, \\ \rho_N(\phi_j, 0; k, x, \omega) = \rho_0(\phi_j; k, x, \omega) \quad j = 1, 2, \ldots, N, \end{cases} \tag{4}$$

where

$$F_N(\phi) = \frac{1}{\overline{P}} \sum_{j'=1}^{N} \int_{\mathbf{R}} \delta G(x-y)\mathrm{d}y \int_{\mathbf{R}} g(\omega')\mathrm{d}\omega' \int_{\mathbf{R}_+} k' P(k')\mathrm{d}k'$$
$$\times \Gamma(\phi - \phi_{j'})\rho_N(\phi_{j'}, t - \tau(x, y); k', y, \omega').$$

Note that this function is defined on a continuous interval with respect to ϕ. By virtue of the estimate (9), which we will verify later, it is clear that problem (4) above has a unique solution ρ_N for every $N \in \mathbf{N}$. Then, we derive some bounds for ρ_N and its differences, which are uniform with respect to N. To do that, in the space of grid functions, we define the scalar product and the norm by

$$(f_1, f_2)_\delta \equiv \sum_{j=1}^{N} f_1(\phi_j)\overline{f_2(\phi_j)}\delta, \quad \|f\|_\delta^2 \equiv (f, f)_\delta,$$

respectively. As Sjöberg [12] and Tsutsumi [14] did, we assume $N = 2n + 1$ with $n \in \mathbf{N}$, and then

$$\left\{\frac{1}{\sqrt{2\pi}}\mathrm{e}^{\mathrm{i}l\phi}\right\}_{l=-n}^{n}$$

forms an orthonormal basis with respect to the scalar products $(\cdot, \cdot)_\delta$ and $(\cdot, \cdot)$.

The following lemmas are due to the works of Sjöberg [12] and Tsutsumi [14]; thus, we omit the proof here.

Lemma 1 *If f_1 is a real N-periodic grid-function, i.e., if $f_1(\phi_j) = f_1(\phi_{j+N})$ ($j = 1, 2, \ldots, N$), and if $f_2(\phi)$ is another real N-periodic grid-function, then the following equalities hold:*

$$(f_1, \partial_+ f_2)_\delta = -(\partial_- f_1, f_2)_\delta, \quad (f_1, \partial_- f_2)_\delta = -(\partial_+ f_1, f_2)_\delta, \tag{5}$$

$$(\partial_+ f_1, f_1)_\delta = -\frac{\delta}{2}\|\partial_+ f_1\|_\delta^2. \tag{6}$$

Lemma 2 *Let q_1 and q_2 be non-negative integers with $q_1 + q_2 = q$, and ψ, a function of the form*

$$\psi(\phi) = \frac{1}{\sqrt{2\pi}} \sum_{l=-n}^{n} a_l \mathrm{e}^{\mathrm{i}l\phi}.$$

Then,

$$c_{41} \left\| \frac{\partial^q \psi}{\partial \phi^q} \right\|^2 \leq \|\partial_+^{q_1} \partial_-^{q_2} \psi\|^2 = \|\partial_+^{q_1} \partial_-^{q_2} \psi\|_\delta^2 \leq c_{42} \left\| \frac{\partial^q \psi}{\partial \phi^q} \right\|^2$$

holds with some constants c_{4j} $(j = 1, 2)$.

Lemma 3 *Let $f = f(\phi)$, and ψ be the discrete Fourier series of f, that is,*

$$\psi(\phi) = \frac{1}{\sqrt{2\pi}} \sum_{l=-n}^{n} b_l \mathrm{e}^{\mathrm{i}l\phi}$$

with $b_l = \frac{1}{\sqrt{2\pi}}(\mathrm{e}^{\mathrm{i}l\phi}, f)_\delta$. Then, the discrete version of the Parseval-type equality

$$\|\partial_+^{q_1} \partial_-^{q_2} \psi\|_\delta^2 = \|\partial_+^{q_1} \partial_-^{q_2} f\|_\delta^2$$

holds for non-negative integers q_1 and q_2.

On the basis of Lemmas 1–3, we derive some estimates of ρ_N and its differences.

Lemma 4 *The following estimates hold:*

$$\begin{aligned}
\|\rho_N(\cdot, t; k, x, \omega)\|_\delta &\leq c_{43} \quad \forall t > 0, \ (k, x, \omega) \in \mathbf{R}_+ \times \mathbf{R}^2, \\
\|\partial_- \rho_N(\cdot, t; k, x, \omega)\|_\delta &\leq c_{44} \quad \forall t > 0, \ (k, x, \omega) \in \mathbf{R}_+ \times \mathbf{R}^2, \\
\|\partial_+^j \partial_-^j \rho_N(\cdot, t; k, x, \omega)\|_\delta &\leq \tilde{c}_{4(j)} \quad j = 1, 2, \ldots, m, \ \forall t > 0, \ (k, x, \omega) \in \mathbf{R}_+ \times \mathbf{R}^2,
\end{aligned}$$

where $c_{4(l)}$ $(l = 3, 4)$ and $\tilde{c}_{4(j)}$ $(j = 1, 2)$ are positive constants independent of t, k, x, and ω.

Proof Let us multiply $(4)_1$ by ρ_N. Then, by virtue of Lemmas 1–2, we have

$$\begin{aligned}
&\frac{1}{2}\Big\{\|\rho_N(\cdot, T; k, x, \omega)\|_\delta^2 - \|\rho_N(\cdot, 0; k, x, \omega)\|_\delta^2\Big\} + \frac{\omega\delta}{2} \int_0^T \|\partial_+ \rho_N(\cdot, t; k, x, \omega)\|_\delta^2 \, \mathrm{d}t \\
&\quad + D \int_0^T \|\partial_- \rho_N(\cdot, t; k, x, \omega)\|_\delta^2 \, \mathrm{d}t \\
&\qquad \leq k \int_0^T \left| \Big(\partial_-(F_N \rho_N), \rho_N\Big)_\delta \right| \mathrm{d}t \qquad (7)
\end{aligned}$$

for arbitrary $T > 0$. Making use of

$$\partial_-(f_1 f_2) = f_2(\partial_- f_1) + (f_1)_- \partial_- f_2$$

for two real N-periodic grid functions f_1 and f_2 in general, where $(f_1)_-(\phi_j) = f_1(\phi_{j-1})$, we estimate the right-hand side of (7). Let us show a part of the estimates in the following. By virtue of the variable transform and the Schwartz's inequality, we have

$$\begin{aligned}
|F_N| &= \frac{1}{P}\Bigg|\sum_{j'=1}^{N}\int_0^\infty \delta G(x-y(x,\tau))\Big(\frac{\partial\tau}{\partial y}\Big)^{-1}\mathrm{d}\tau\int_{\mathbf{R}} g(\omega')\mathrm{d}\omega' \\
&\qquad \times \int_{\mathbf{R}_+} k'P(k')\Gamma(\phi-\phi_{j'})\rho_N(\phi_{j'}, t-\tau; k', y, \omega')\mathrm{d}k'\Bigg| \\
&= \frac{1}{P}\Bigg|\sum_{j'=1}^{N}\int_0^t \delta G(x-y(x,\tau))\Big(\frac{\partial\tau}{\partial y}\Big)^{-1}\mathrm{d}\tau\int_{\mathbf{R}} g(\omega')\mathrm{d}\omega' \\
&\qquad \times \int_{\mathbf{R}_+} k'P(k')\Gamma(\phi-\phi_{j'})\rho_N(\phi_{j'}, t-\tau; k', y, \omega')\mathrm{d}k'\Bigg| + c_{45} \\
&\le c_{46}\Big(\int_0^t \Big|G(x-y(x,\tau))\Big|^{q'}\mathrm{d}\tau\Big)^{1/q'}\Big(\int_0^t \|\rho_N(t-\tau)\|_\delta^q \mathrm{d}\tau\Big)^{1/q} + c_{45},
\end{aligned}$$

with $1/q + 1/q' = 1$. Thereby, we have

$$\begin{aligned}
&\int_0^T \Big|\big(F_N\partial_-\rho_N, \rho_N\big)_\delta\Big|\,\mathrm{d}t \\
&\qquad \le \int_0^T c_{47}(t)\Big(\int_0^t \|\rho_N(t-\tau)\|_\delta^q \mathrm{d}\tau\Big)^{1/q}\Big\{\|\partial_-\rho_N(t)\|_\delta^p + \|\rho_N(t)\|_\delta^{p'}\Big\}\,\mathrm{d}t \\
&\qquad\qquad + c_{45}\int_0^T \Big\{\varepsilon_1\|\partial_-\rho_N(t)\|_\delta^2 + C_{\varepsilon_1}\|\rho_N(t)\|_\delta^2\Big\}\,\mathrm{d}t,
\end{aligned}$$

where ε_1 is a certain positive constant, and C_{ε_1}, a constant dependent on ε_1 (hereafter we use these notations in the same meaning). $c_{47}(t)$ is a non-negative and non-decreasing function of its argument. Then, by taking $pq' = 2$ and applying the Young's inequality, we have

$$\begin{aligned}
&\int_0^T \frac{c_{47}(t)}{2}\Big(\int_0^t \|\rho_N(t-\tau)\|_\delta^q \mathrm{d}\tau\Big)^{1/q}\|\partial_-\rho_N(t)\|_\delta^p\,\mathrm{d}t \\
&\qquad \le \varepsilon_1\int_0^T \|\partial_-\rho_N(t)\|_\delta^2\,\mathrm{d}t + C_{\varepsilon_1}T\int_0^T \|\rho_N(t)\|_\delta^q \mathrm{d}t,
\end{aligned}$$

$$\int_0^T \frac{c_{47}(t)}{2}\left(\int_0^t \|\rho_N(t-\tau)\|_\delta^q \mathrm{d}\tau\right)^{1/q} \|\rho_N(t)\|_\delta^{p'} \,\mathrm{d}t$$
$$\leq \frac{c_{47}(T)T}{4}\int_0^T \|\rho_N(\tau)\|_\delta^q \mathrm{d}\tau + \frac{c_{47}(T)}{4}\int_0^T \|\rho_N(t)\|_\delta^{p'q'} \,\mathrm{d}t.$$

Thus, we have

$$\int_0^T \left|\left(F_N \partial_- \rho_N, \rho_N\right)_\delta\right| \,\mathrm{d}t \leq c_{48}(T)\int_0^T \left\{\|\rho_N(t)\|_\delta^q + \|\rho_N(t)\|_\delta^{p'q'} + \|\rho_N(t)\|_\delta^2\right\} \mathrm{d}t$$
$$+ \left(\varepsilon_2 + c_{45}\varepsilon_1\right)\int_0^T \|\partial_- \rho_N(t)\|_\delta^2 \,\mathrm{d}t.$$

We finally have the estimate of the form

$$\|\rho_N(T)\|_\delta^2 \leq c_{49} + c_{410}(T)k\int_0^T w\left(\|\rho_N(t)\|_\delta^2\right) \mathrm{d}t \tag{8}$$

with sufficiently regular, positive non-decreasing $c_{410}(T)$ and $w(x)$, satisfying $w(x) \geq Lx$ with a certain $L > 0$ and $w(0) = 0$ (take, for instance, $(q, q') = (3, 3/2)$, $(p, p') = (4/3, 4)$.) Now, we introduce a notation

$$\|\rho_N(t)\|_{\delta(\alpha_1)}^2 \equiv \sup_{k,x,\omega} \mathrm{e}^{-\alpha_1 kt}\|\rho_N(\cdot, t; k, x, \omega)\|_\delta^2.$$

By virtue of the Constantin's extended version of the Gronwall-Bellman-Bihari's inequality [3, 4], we then have the estimate of the form

$$|||\rho_N(t)|||_{\delta(\alpha_1)}^2 \leq c_{411} \quad t \in (0, T_*) \tag{9}$$

on a certain time interval $(0, T_*)$ with a sufficiently large α_1 and a constant c_{411} independent of k, x, and ω. The second and third inequalities in the statement of Lemma 4 are obtained by multiplying $(4)_1$ by $\partial_+\partial_-\rho_N$ and $\partial_+^2\partial_-^2\rho_N$, respectively, and then applying the similar but more lengthy calculations. We omit detailed discussion here due to space limitations. □

By virtue of Lemma 4, we easily obtain

$$\left\|\frac{\mathrm{d}\rho_N}{\mathrm{d}t}(t)\right\|_{\delta(\alpha_1)} \leq c_{412}, \quad \left\|\partial_+^2\partial_-^2\frac{\mathrm{d}\rho_N}{\mathrm{d}t}(t)\right\|_{\delta(\alpha_1)} \leq c_{413} \tag{10}$$

under the assumptions of Theorem 1. Then, as Sjöberg [12] and Tsutsumi [14] did, we consider the discrete Fourier series of ρ_N, which is denoted as $\{\psi_N\}$:

$$\psi_N(\phi, t; k, x, \omega) = \frac{1}{\sqrt{2\pi}}\sum_{l=-n}^{n} a_l(t; k, x, \omega)\mathrm{e}^{\mathrm{i}l\phi}, \quad a_l(t; k, x, \omega) = \frac{1}{\sqrt{2\pi}}\left(\mathrm{e}^{\mathrm{i}l\phi}, \rho_N\right)_\delta.$$

Estimate (10) and Lemmas 3–4 yield that the sequence of functions $\{\psi_N\}$ is uniformly bounded and equicontinuous on $0 \le \phi \le 2\pi, 0 \le t \le T$. With the aid of the Arzera-Ascoli theorem, we see that $\{\psi_N\}$ contains a subsequence that converges to a certain function ρ as $N \to +\infty$. In addition, it is clear that

$$\left(\partial/\partial\phi\right)^m \psi_N \to \left(\partial/\partial\phi\right)^m \rho \quad (N \to +\infty)\ (m = 1, 2, \ldots, 4)$$

in $\overline{\mathscr{H}^0}$. Therefore, this ρ is the desired solution to (3). This guarantees the existence of a solution to (3).

4.2 Proof of $\rho \in \mathscr{V}^4_{(\alpha_1, \alpha_2)}(T_*)$

Here, we prove that the solution ρ guaranteed in the previous process certainly meets the natural characteristics of the probability density. We also use a notation before:

$$F[\rho_1, \rho_2] \equiv \frac{1}{\overline{P}}\rho_1(\phi, t; k, x, \omega) \int_{\mathbf{R}} G(x-y)\mathrm{d}y \int_{\mathbf{R}} g(\omega')\mathrm{d}\omega' \int_{\mathbf{R}_+} k' P(k')\mathrm{d}k' \\ \int_0^{2\pi} \Gamma(\phi - \phi')\rho_2(\phi, t - \tau(x, y); k', y, \omega')\, \mathrm{d}\phi'.$$

Lemma 5 *For an arbitrary $T > 0$, if there exists a solution ρ to (3) that belongs to $\mathscr{V}^4_{(\alpha_1, \alpha_2)(*)}(T)$, then the following items hold:*

(i) $\displaystyle\int_\Omega \rho(\phi, t; k, x, \omega)\, \mathrm{d}\phi = 1 \quad t \in (0, T),\ (k, x, \omega) \in \mathbf{R}_+ \times \mathbf{R}^2;$

(ii) $\rho(\phi, t; k, x, \omega) \ge 0$ a.e. on $\phi \in \Omega,\ t \in (0, T),\ (k, x, \omega) \in \mathbf{R}_+ \times \mathbf{R}^2$.

Proof The first statement is proved by direct calculation (see, [7]), and we omit the proof. We verify the second statement by using the Stampacchia's truncation method. Let us define

$$\rho_+ \equiv (|\rho| + \rho)/2 \ge 0, \quad \rho_- \equiv (|\rho| - \rho)/2 \ge 0.$$

It is obvious that ρ_+ and ρ_- stand for the positive and negative parts of ρ, respectively, satisfying $\rho = \rho_+ - \rho_-$. Then, by multiplying $(3)_1$ by ρ_- and integrating over Ω, we obtain

$$\frac{1}{2}\frac{\mathrm{d}}{\mathrm{d}t}\|\rho_-(t; k, x, \omega)\|^2 + D\|\rho_-^{(1,0)}(t; k, x, \omega)\|^2 \le kc_{414}\|\rho_-(t; k, x, \omega)\|^2.$$

Here, we used the estimate:

$$\int_\Omega \rho_-(\phi, t; k, x, \omega)\frac{\partial}{\partial\phi}\left(F[\rho, \rho]\right)\, \mathrm{d}\phi \le \frac{C_1 M_1}{2}\|\rho_-(\cdot, t; k, x, \omega)\|^2,$$

which is derived by integration by parts. Taking into account $\rho_-\big|_{t=0} = 0$, we arrive at $\rho_- \equiv 0$ on $t \in (0, T)$ by virtue of the Gronwall's inequality. This implies the second statement.

Therefore, the solution that was guaranteed to exist in the previous process also belongs to $L_1^{(1)}(T_*)$ thanks to Lemma 5 and consequently to $\mathscr{V}^4_{(x_1,x_2)}(T_*)$.

4.3 Uniqueness of Solution

Finally, we discuss the uniqueness part of the statement. Assume that there exist two solutions ρ_i $(i = 1, 2)$ to (3) on $(0, T_*)$ associated with the same initial data, and let us define $\tilde{\tilde{\rho}} \equiv \rho_1 - \rho_2$. Then, it satisfies

$$\begin{cases} \dfrac{\partial \tilde{\tilde{\rho}}}{\partial t} + \omega \dfrac{\partial \tilde{\tilde{\rho}}}{\partial \phi} - D \dfrac{\partial^2 \tilde{\tilde{\rho}}}{\partial \phi^2} + k \dfrac{\partial}{\partial \phi}\Big(F[\tilde{\tilde{\rho}}, \rho_1]\Big) + k \dfrac{\partial}{\partial \phi}\Big(F[\rho_2, \tilde{\tilde{\rho}}]\Big) = 0 \\ \qquad\qquad \phi \in \Omega,\ t \in (0, T_*),\ (k, x, \omega) \in \mathbf{R}_+ \times \mathbf{R}^2, \\ \dfrac{\partial^j \tilde{\tilde{\rho}}}{\partial \phi^j}\Big|_{\phi=0} = \dfrac{\partial^j \tilde{\tilde{\rho}}}{\partial \phi^j}\Big|_{\phi=2\pi} \quad (j = 0, 1),\ t \in (0, T_*),\ (k, x, \omega) \in \mathbf{R}_+ \times \mathbf{R}^2, \\ \tilde{\tilde{\rho}}|_{t=0} = 0 \quad \phi \in \Omega,\ (k, x, \omega) \in \mathbf{R}_+ \times \mathbf{R}^2. \end{cases} \tag{11}$$

We multiply $(11)_1$ by $\tilde{\tilde{\rho}}$. Then, integration by parts yields

$$\frac{1}{2}\frac{\mathrm{d}}{\mathrm{d}t}\|\tilde{\tilde{\rho}}(t; k, x, \omega)\|^2 + D\|\tilde{\tilde{\rho}}^{(1,0)}(t; k, x, \omega)\|^2 \le kC_1M_1\|\tilde{\tilde{\rho}}(t; k, x, \omega)\|^2 \quad \forall t \in (0, T_*).$$

By virtue of the Gronwall's inequality and the initial condition $(1)_3$, we have

$$|||\tilde{\tilde{\rho}}(t; k, x, \omega)|||^2 = 0 \quad \forall t > 0.$$

This completes the uniqueness and consequently the proof of Theorem 1.

5 Conclusion and Remaining Issues

In this paper, we introduced a functional equation-based model of the resting-state network and showed its local-in-time solvability. The global-in-time solvability, stability analysis, and bifurcation analysis are to be studied in the near future.

References

1. Cabral, J., Hugues, E., Sporns, O., Deco, G.: Role of local network oscillations in resting-state functional connectivity. NeuroImage **57**, 130–139 (2011)
2. Cabral, J., Kringelbach, M.L., Deco, G.: Exploring the network dynamics underlying brain activity during rest. Prog Neurobiol. **114**, 102–131 (2014)
3. Constantin, A.: A Gronwall-like inequality and its applications. Rend. Mat. Acc. Lincei **9**, 111–115 (1990)
4. Constantin, A.: On some integro-differential and integral inequalities and applications, Ann. Univ. Din Timis. Facult. de Mat. Infor. **30**, 1–21 (1990)
5. Deco, G., Jirsa, V., McIntosh, A.R., Kötter, R.: Key role of coupling, delay and noise in resting brain fluctuations. Proc. Natl. Acad. Sci. U.S.A. **106**, 10302–10307 (2009)
6. Ichinomiya, T.: Frequency synchronization in a random oscillator network. Phys. Rev. E **70**, 026116 (2004)
7. Lavrentiev, M., Spigler, R.S.: Existence and uniqueness of solutions to the Kuramoto-Sakaguchi nonlinear parabolic integrodifferential equation. Differ. Integr. Equ. **13**, 649–667 (2000)
8. Lee, W.S., Odd, E., Antonsen, T.M.: Large coupled oscillator systems with heterogeneous interaction delays. Phys. Rev. Lett. **103**, 044101 (2009)
9. Moussa, M.N., Steen, M.R., Laurienti, P.J., Hayasaka, S.: Consistency of network modules in resting-state fMRI connectome data. PLOS One **7**, e44428 (2012)
10. Gusnard, D.A., Raichle, M.E.: Searching for a baseline: functional imaging and the resting human brain. Nat. Rev. Neurosci. **2**, 685–694 (2001)
11. Raichle, M.E., MacLeod, A.M., Snyder, A.Z., Powers, W.J., Gusnard, D.A., Shulman, G.L.: A default mode of brain function. Proc. Natl. Acad. Sci. USA, **98**, 676–682
12. Sjöberg, A.: On the Korteweg-de Vries equation. J. Math. Anal. Appl. **29**, 569–579 (1970)
13. Temam, R.: Infinite-Dimensional Dynamical Systems in Mechanics and Physics, 2nd (edn.). Springer, New York (1997)
14. Tsutsumi, M., Mukasa, T.: Parabolic regularizations for the generalized korteweg-de vries equation. Funkcialaj Ekvacioj **14**, 89–110 (1971)
15. Wilson, H.R., Cowan, J.D.: A mathematical theory of the functional dynamics of cortical and thalamic nervous tissue. Kybernetik **13**(2), 55–80 (1973)

Convergence Rates for Discrete-to-Continuum Limits in 1D Particle Systems

Patrick van Meurs

Abstract We contribute to a recent series of papers on discrete-to-continuum limits of one-dimensional particle systems governed by nonlocal and unbounded interactions. While convergence of the equilibrium positions of the particles to a limiting density profile is known, and while in several cases boundary layers have been characterised, any quantitative bound on the convergence rate to the limiting density is missing. Our main result guarantees such bounds. The proof method relies on quantitative versions of the compactness estimates in a recent paper on boundary layers, supplemented with new estimates, which give a precise meaning to the deviation from the particle positions in equilibrium to the equispaced configuration.

1 Introduction

The starting point of our analysis is the minimisation problem of the energy E_n over the domain $\mathscr{D}_n$, which are defined by

$$\mathscr{D}_n := \{x \in [0,1]^{n+1} : 0 =: x_0 \leq x_1 \leq \ldots \leq x_n := 1\};$$
$$E_n : \mathscr{D}_n \to \mathbb{R}, \quad E_n(x) := \frac{1}{n} \sum_{k=1}^{n} \sum_{j=0}^{n-k} \alpha_n V\big(n\alpha_n(x_{k+j} - x_j)\big). \tag{1}$$

We interpret $n \in \mathbb{N}$ as the number of particles, and the unknowns x_i as the particle positions in $[0,1]$ (see Fig. 1a for a typical configuration $x \in \mathscr{D}_n$). $V : (0,\infty) \to (0,\infty)$ is the interaction potential (Fig. 1b illustrates a typical example, and in (3), we list the precise properties of V), which models an unbounded, nonlocal repulsive force between any two particles. The parameter $\alpha_n > 0$ relates the length-scale of the decay of the tails of V to the length of the domain $[0,1]$. The repulsive interaction

P. van Meurs (✉)
Kanazawa University, Kakuma, Kanazawa 920-1192, Japan
e-mail: pmeurs@staff.kanazawau.ac.jp

P. van Meurs et al. (eds.), *Mathematical Analysis of Continuum Mechanics and Industrial Applications II*, Mathematics for Industry 30,
https://doi.org/10.1007/978-981-10-6283-4_15

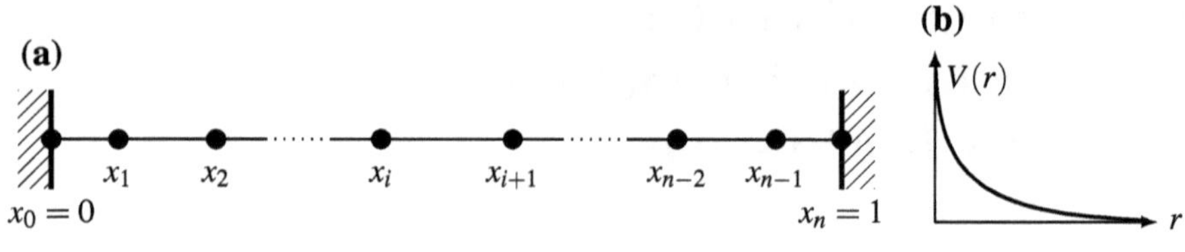

Fig. 1 **a** A typical $x \in \mathscr{D}_n$ with low energy. **b** A typical potential V

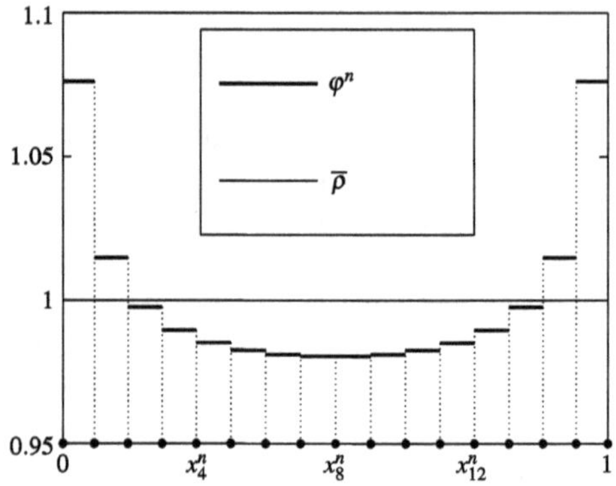

Fig. 2 The minimiser x^n of E_n (1) with $n = 16$, $\alpha_n = 1$ and $V(r) = r^{-2}$. φ^n is computed from x^n; see (6). Theorem 1 provides the rate at which φ^n converges to $\overline{\rho} \equiv 1$ as $n \to \infty$

force and the confinement of the particles to the interval [0, 1] result in a nontrivial minimisation problem of E_n. Figure 2 illustrates the minimiser $x^n \in \mathscr{D}_n$ of E_n for $V(r) = r^{-2}$.

The main aim of this chapter is to study the minimisation of E_n in the limit $n \to \infty$. This limit passage is established in [4] on the half-infinite domain $[0, \infty)$ with an additional potential field which pushes the particles to the end of the domain at 0. In particular, [4, Theorem 1] states Γ-convergence of E_n in the narrow topology (2) to a limit energy E, which depends on a particle density $\rho \in \mathscr{P}([0, 1])$, where $\mathscr{P}([0, 1])$ is the space of probability measures. In particular, this means that the empirical measure of the minimiser x^n, given by $\mu^n := \frac{1}{n} \sum_{i=0}^n \delta_{x_i^n}$, converges to the minimiser $\overline{\rho}$ of E in the narrow topology, i.e.

$$\int_0^1 \varphi \, d\mu^n \xrightarrow{n \to \infty} \int_0^1 \varphi \overline{\rho} \quad \text{for all } \varphi \in C([0, 1]), \tag{2}$$

which we denote by $\mu^n \rightharpoonup \overline{\rho}$. The energy E and its minimiser $\overline{\rho}$ depend on the asymptotic scaling of α_n; in the critical cases where $n\alpha_n \to \alpha$ or $\alpha_n \to \alpha$ for some $\alpha > 0$ as $n \to \infty$, E and $\overline{\rho}$ depend on α. For all other asymptotic behaviour of α_n separated by these two critical regimes (for instance, $\alpha_n = 1/\sqrt{n}$ or $\alpha_n = n$), three other energies E are found. Since E_n models pile-ups of n dislocation walls, the limiting energies E are models for the dislocation density (used for describing plasticity of metals). Indeed, for several choices of (α_n) different models in the engineering literature are captured by E. However, while Γ-convergence implies that the minimisers x^n of E_n converge to $\overline{\rho}$, it remains unknown for given data n and α_n which of the limiting profiles $\overline{\rho}$ fits best.

Theorem 1, our main result, contributes to this open problem by providing the first result on a quantitative bound (the *convergence rate*) on the difference between x^n and $\overline{\rho}$.

Theorem 1 fits within the series of papers starting with [4]. In [12], the Γ-convergence result of E_n in [4] is extended to bounded domains, which captures (1). [11] establishes evolutionary convergence of the gradient flows of E_n. The papers [3, 5–7] characterise boundary layers in the profile of x^n close to the endpoints 0 and 1. A common observation in numerical computations of x^n in all these papers is that φ^n and $\overline{\rho}$ show a remarkably good resemblance, while the current Γ-convergence results on E_n only guarantee a much weaker notion of convergence. Theorem 1 is the first rigorous result on this resemblance.

Moreover, Theorem 1 covers potentials V for which the convergence of minimisers as in (2) as well as Γ-convergence of E_n is not known yet. In particular, for $\frac{1}{n}\alpha_n \to 0$, Theorem 1 allows for potentials V which have a nonintegrable singularity (see Sect. 4.2 for examples). Such potentials are studied, for instance, in approximation theory [2, 9], where convergence of the minimisers is known on periodic domains (as opposed to the domain $\mathscr{D}_n$).

The remainder of this chapter is organised as follows. Section 2 defines the framework to state and prove Theorem 1, which is done in Sect. 3. In Sect. 4, we apply Theorem 1 to obtain asymptotic bounds on α_n for which the distance between φ^n and $\overline{\rho}$ vanishes in the limit. We conclude in Sect. 5.

2 Framework and Notation

We set $\mathbb{N} := \{1, 2, \ldots\}$ and consider E_n as in (1). The precise conditions on the interaction potential V are similar to those in [7]:

$$V \in C^2((0,\infty)) \text{ is positive;} \tag{3a}$$

$$V(r) \xrightarrow{r\to 0} \infty \text{ and } V(r) \xrightarrow{r\to\infty} 0; \tag{3b}$$

$$V'' > 0 \text{ is nonincreasing on } (0,\infty); \tag{3c}$$

$$\Sigma := \int_{\frac{1}{2}}^{\infty} U(r)^2\,dr < \infty, \text{ where } U(r) := \int_r^{\infty} V(s)\,ds. \tag{3d}$$

From (3), we conclude that $V^{-1} : (0,\infty) \to (0,\infty)$ exists. Equation (3c) is chosen for convenience; it can be weakened to V'' being bounded from below on $(0,\infty)$ by a positive, nonincreasing function (as is done in [7]). Equation (3d) is an assumption on the tail of V, which is weaker than the assumption [7, **(Dec+)**].

From [10, Sect. 3.6] and [12], it follows that $\inf_n \alpha_n > 0$ implies

$$\mu^n \rightharpoonup \overline{\rho} \equiv 1 \quad \text{as } n \to \infty. \tag{4}$$

Moreover, if V is integrable, then also the case $n\alpha_n \to \infty$ as $n \to \infty$ implies (4). In [7], boundary layers are studied by means of an expansion of E_n, whose first-order term reads

$$E_n^1(x) := n\Big[E_n(x) - \frac{n+1-k}{n}\sum_{k=1}^{n} V(k)\Big] = \sum_{k=1}^{n}\sum_{j=0}^{n-k}\big[V\big(n(x_{k+j}-x_j)\big) - V(k)\big]. \tag{5}$$

To detect boundary layers in the minimiser x^n, [7] changes variables to

$$\varepsilon^n \in \mathbb{R}^n, \quad \varepsilon_i^n := n(x_i^n - x_{i-1}^n) - 1 \text{ for all } i = 1, \dots, n.$$

Then, φ^n as illustrated in Fig. 2 reads

$$\varphi^n \in L^\infty(0,1), \quad \varphi^n(r) := \frac{1/n}{x_i^n - x_{i-1}^n} = \frac{1}{\varepsilon_i^n + 1} \quad \text{with } i \text{ such that } x_{i-1}^n < r < x_i^n. \tag{6}$$

We measure the difference between φ_n and $\overline{\rho}$ in both the L^2-norm and the q-Wasserstein distance for any $q \geq 1$. The L^2-norm is a natural choice given Fig. 2, and the q-Wasserstein distance metrises the narrow topology (2) for any $q \geq 1$ [1, Proposition 7.1.5]. We define the q-Wasserstein distance by its characterising property in 1D given by

$$W_q(\mu_1, \mu_2) := \Big(\int_0^1 |\xi_1(t) - \xi_2(t)|^q \, dt\Big)^{\frac{1}{q}},$$

where $\mu_1, \mu_2 \in \mathscr{P}([0,1])$, and $\xi_i : [0,1] \to [0,1]$ are nondecreasing functions which are related to μ_i by

$$(\xi_i^{-1})' = \mu_i \quad \text{in } \mathscr{D}'(0,1), \tag{7}$$

where we make use of the pseudo-inverse.[1] In particular, interpreting $\overline{\rho}, \varphi^n \in \mathscr{P}([0,1])$, we obtain, respectively, from (7) that

$$\overline{\xi}(t) = t, \quad \text{and } \xi^n \text{ is the linear interpolation of } \big((\tfrac{i}{n}, x_i^n)\big)_{i=0}^n \subset \mathbb{R}^2.$$

[1] We need the *pseudo*-inverse in case ξ_i contains jumps or flat parts.

Next, we show that the L^2-distance is stronger than the q-Wasserstein distance for any q. Indeed,

$$\begin{aligned} W_q(\varphi^n, \overline{\rho}) &= \|\xi^n - \overline{\xi}\|_{L^q(0,1)} \le \|\xi^n - \overline{\xi}\|_{L^\infty} \le \|(\xi^n)' - (\overline{\xi})'\|_{L^1} \\ &= \int_0^1 \left| 1 - \frac{1}{(\xi^n)'(t)} \right| (\xi^n)'(t)\, dt = \int_0^1 \left| 1 - \frac{1}{(\xi^n)'\big((\xi^n)^{-1}(r)\big)} \right| dr \\ &= \int_0^1 |1 - \varphi^n(r)|\, dr = \|\varphi^n - \overline{\rho}\|_{L^1} \le \|\varphi^n - \overline{\rho}\|_{L^2}. \end{aligned}$$

This difference in topologies is apparent from the following thought experiment. If x_i^n is allowed to vary, then $W_q(\varphi^n, \overline{\rho})$ deviates by $\mathcal{O}(\frac{1}{n})$ as x_i^n moves to x_{i-1}^n, while $\|\varphi^n - \overline{\rho}\|_{L^2}$ blows up as x_i^n moves to x_{i-1}^n.

3 Convergence Rate: Statement and Proof

Theorem 1 (Convergence rate) *Let $\alpha_n = 1$, and let Σ, $\overline{\rho}$ and φ^n be defined as in* (3d), (4) *and* (6), *respectively. We further set*

$$p(r) := r(r+1) \quad \text{and} \quad \Lambda := U(\tfrac{1}{2}) + \Sigma / V''(2). \tag{8}$$

Then, there exist $c_1, c_2 > 0$ such that for any V satisfying (3), any $n \in \mathbb{N}$ and any $q \ge 1$ it holds that

$$\begin{aligned} \|\varphi^n - \overline{\rho}\|_{L^2(0,1)}^2 &\le \frac{\Xi}{n}, & &\text{where } \Xi := \frac{c_2}{V^{-1}(c_1 \Lambda)} p\left(\frac{\Sigma}{V''(2)^2} \right), \\ W_q^2(\varphi^n, \overline{\rho}) &\le \Theta^n \le \frac{\Xi}{n}, & &\text{where } \Theta^n := c_2 p\left(\frac{\Sigma}{n V''(2)^2} \right). \end{aligned}$$

The proof of Theorem 1 relies on a more quantitative version of several estimates in [7] (Lemma 1) and two new estimates (Lemma 2). To state these estimates, we use the following definitions from [7]

$$\begin{aligned} &\phi_k : (-k, \infty) \to [0, \infty), \quad \phi_k(r) := V(k+r) - V(k) - V'(k) r; \\ &\sigma^n \in [0,\infty)^n, \quad \sigma_i^n := \begin{cases} \sum\limits_{k=i+1}^{n-i} \big((k-i) \wedge (n-i+1-k)\big)[-V'(k)] & \text{if } i \le \lfloor n/2 \rfloor, \\ 0 & \text{otherwise}; \end{cases} \\ &\sigma^\infty \in \ell^2(\mathbb{N}), \quad \sigma_i^\infty := \sum_{k=i+1}^{\infty} (k-i)[-V'(k)]; \\ &\mathcal{D}_n^1 := \big\{ \varepsilon \in [-1, n]^n : \exists x \in \mathcal{D}_n \, \forall 1 \le i \le n : \varepsilon_i = n(x_i - x_{i-1}) - 1 \big\}. \end{aligned}$$

Then, the boundary-layer energy $E_n^1 : \mathscr{D}_n^1 \to \mathbb{R}$ defined in (5) reads

$$E_n^1(\varepsilon) = \sum_{k=1}^{n} \sum_{j=0}^{n-k} \phi_k \Big(\sum_{l=j+1}^{k+j} \varepsilon_l \Big) + \sigma^n \cdot (\varepsilon + \varepsilon^{\leftarrow}), \tag{9}$$

where $\varepsilon^{\leftarrow} \in \mathscr{D}_n^1$ is the 'reversed' version of ε, i.e. $\varepsilon_i^{\leftarrow} := \varepsilon_{n+1-i}$. We also set $\varepsilon_{\pm} \in \mathscr{D}_n^1$ as the positive/negative part of ε, i.e. $\varepsilon_{\pm,i} := [\varepsilon_i]_{\pm}$. If necessary, we extend $\varepsilon \in \mathscr{D}_n^1$ to $\ell^2(\mathbb{N})$ by 0-valued entries without changing notation.

For functions $f, g : (0, \infty) \to \mathbb{R}$ we say that $f(r) \lesssim g(r)$ if there exists a universal constant $C > 0$ such that $f(r) \le C g(r)$ for all $r > 0$. We further define $f(r) \lesssim g(r)$ for $r \gg 1$ if there exist universal constants $C_1, C_2 > 0$ such that $f(r) \le C_1 g(r)$ for all $r > C_2$. We use $\gtrsim$ and $\ll$ for the opposite inequalities. Finally, we say that $f(r) \sim g(r)$ if $f(r) \lesssim g(r)$ and $f(r) \gtrsim g(r)$.

Lemma 1 (Quantitative estimates from [7]) *For all $n \in \mathbb{N}$*

$$\phi_1(r) \ge \Phi(r) := \frac{V''(2)}{2} \begin{cases} r^2 & -1 < r \le 1, \\ r & 1 < r; \end{cases} \tag{10}$$

$$\|\sigma^\infty\|_{\ell^2(\mathbb{N})}^2 \le \Sigma; \tag{11}$$

$$0 \ge E_n^1(\varepsilon^n) \gtrsim \sum_{i=1}^{n} \Phi\left(\varepsilon_{+,i}^n\right) + \frac{1}{V''(2)} \left\| V''(2)\varepsilon_-^n - \sigma^\infty \right\|_{\ell^2(\mathbb{N})}^2 - \frac{\Sigma}{V''(2)}. \tag{12}$$

Proof From the minimality of ε^n and (9), we obtain $E_n^1(\varepsilon^n) \le E_n^1(0) = 0$. The other inequalities in (10) and (12) are given by [7, Lemma 4.1 and (65)], respectively. Equation (11) follows from recognising a Riemann lower-sum in

$$\sigma_i^\infty = \sum_{k=i+1}^{\infty} (k-i)[-V'(k)] \le \int_i^\infty (r-i)[-V'(r)]\,dr = \int_i^\infty V(r)\,dr = U(i),$$

and using convexity of U^2 to estimate

$$\|\sigma^\infty\|_{\ell^2(\mathbb{N})}^2 = \sum_{i=1}^{\infty} (\sigma_i^\infty)^2 \le \sum_{i=1}^{\infty} U(i)^2 \le \int_{\frac{1}{2}}^{\infty} U^2 = \Sigma.$$

Lemma 2 (New estimates) *For all $n \in \mathbb{N}$*

$$\|\varepsilon^n\|_2^2 \lesssim p\big(\Sigma / V''(2)^2\big); \tag{13}$$

$$\|\varepsilon^n\|_1^2 \lesssim n^2 p\big(\tfrac{1}{n}\Sigma / V''(2)^2\big); \tag{14}$$

$$\inf\left\{\varepsilon_i^n + 1 : n \in \mathbb{N},\ i \in \{1, \dots, n\}\right\} \gtrsim V^{-1}(\Lambda). \tag{15}$$

Proof Using that $\Phi \geq 0$ and $E_n^1(\varepsilon^n) \leq 0$, we obtain from (12) that

$$\Big| V''(2)\|\varepsilon_-^n\|_2 - \|\sigma^\infty\|_{\ell^2(\mathbb{N})} \Big|^2 \leq \big\| V''(2)\varepsilon_-^n - \sigma^\infty \big\|_{\ell^2(\mathbb{N})}^2 \lesssim \Sigma.$$

Together with (11), we obtain $\|\varepsilon_-^n\|_2^2 \lesssim \Sigma/V''(2)^2$. To estimate $\|\varepsilon_+^n\|_2^2$, we use that (12) implies $\sum_{i=1}^n \Phi(\varepsilon_{+,i}^n) \lesssim \Sigma/V''(2)$ to deduce

$$\begin{aligned}
\|\varepsilon_+^n\|_2^2 &= \sum_{i\,:\,\varepsilon_{+,i}^n \leq 1} (\varepsilon_{+,i}^n)^2 + \sum_{i\,:\,\varepsilon_{+,i}^n > 1} (\varepsilon_{+,i}^n)^2 \\
&= \sum_{i\,:\,\varepsilon_{+,i}^n \leq 1} \frac{2}{V''(2)} \Phi(\varepsilon_{+,i}^n) + \sum_{i\,:\,\varepsilon_{+,i}^n > 1} \Big(\frac{2}{V''(2)} \Phi(\varepsilon_{+,i}^n) \Big)^2 \\
&\leq \frac{2}{V''(2)} \sum_{i=1}^n \Phi(\varepsilon_{+,i}^n) + \frac{4}{V''(2)^2} \Big[\sum_{i=1}^n \Phi(\varepsilon_{+,i}^n) \Big]^2 \leq p\Big(\frac{2\Sigma}{V''(2)^2} \Big).
\end{aligned} \tag{16}$$

Finally, (13) follows from $\|\varepsilon^n\|_2^2 = \|\varepsilon_+^n\|_2^2 + \|\varepsilon_-^n\|_2^2$.

Next, we prove (14). By using that

$$\sum_{i=1}^n \sqrt{\Phi(\varepsilon_{+,i}^n)} \leq \sqrt{n \sum_{i=1}^n \Phi(\varepsilon_{+,i}^n)},$$

we obtain from a similar estimate as in (16) that

$$\|\varepsilon_+^n\|_1 = \sum_{i\,:\,\varepsilon_{+,i}^n \leq 1} \varepsilon_{+,i}^n + \sum_{i\,:\,\varepsilon_{+,i}^n > 1} \varepsilon_{+,i}^n \lesssim \frac{\sqrt{n\Sigma}}{V''(2)} + \frac{\Sigma}{V''(2)^2}. \tag{17}$$

Since $0 = \sum_{i=1}^n \varepsilon_i^n = \|\varepsilon_+^n\|_1 - \|\varepsilon_-^n\|_1$, we obtain (14) from (17) and $\|\varepsilon^n\|_1 = \|\varepsilon_+^n\|_1 + \|\varepsilon_-^n\|_1 = 2\|\varepsilon_+^n\|_1$.

Next, we prove (15). Since (15) is obvious for $n = 1$, we assume $n \geq 2$. We fix any $i \in \arg\min\{\varepsilon_i^n + 1 : i \in \{1, \ldots, n\}\}$, and set $x \in \mathscr{D}_{n-1}$ as the vector obtained by removing x_i^n from x^n, i.e. $x := (x_0^n, x_1^n, \ldots, x_{i-1}^n, x_{i+1}^n, \ldots, x_n^n)$. Then, by estimating in (5) all interactions with x_i^n from below by 0 except for $V(n(x_i^n - x_{i-1}^n))$, we obtain

$$\begin{aligned}
0 \geq E_n^1(x^n) &\geq \sum_{k=1}^{n-1} \sum_{j=0}^{n-1-k} \big[V\big(n(x_{k+j} - x_j)\big) - V(k) \big] - \sum_{k=1}^n V(k) + V\big(n(x_i^n - x_{i-1}^n)\big) \\
&= \sum_{k=1}^{n-1} \sum_{j=0}^{n-1-k} \big[V_n\big((n-1)(x_{k+j} - x_j)\big) - V_n(k) \big] + \sum_{k=1}^{n-1} (n-k)\big(V_n(k) - V(k)\big) \\
&\qquad - \sum_{k=1}^n V(k) + V(\varepsilon_i^n + 1),
\end{aligned} \tag{18}$$

where $V_n(x) := V(\frac{n}{n-1}x)$.

We use (18) to bound $V(\varepsilon_i^n + 1)$ from above by the other three terms in the right-hand side of (18). By convexity of V, we obtain

$$\sum_{k=1}^{n} V(k) = \sum_{k=1}^{n} V\Big(\int_{k-\frac{1}{2}}^{k+\frac{1}{2}} r\,dr\Big) \leq \sum_{k=1}^{n} \int_{k-\frac{1}{2}}^{k+\frac{1}{2}} V(r)\,dr \leq \int_{\frac{1}{2}}^{\infty} V(r)\,dr = U(\tfrac{1}{2})$$

and

$$\begin{aligned}\sum_{k=1}^{n-1}(n-k)\big(V_n(k)-V(k)\big) &= \sum_{k=1}^{n-1}(n-k)\big(V(k+\tfrac{k}{n-1}) - V(k)\big)\\ &\geq \sum_{k=1}^{n}(n-1)\frac{k}{n-1}V'(k) = \sum_{k=1}^{n}\sum_{j=1}^{k} V'(k) = \sum_{j=1}^{n}\sum_{k=j}^{n} V'(k).\end{aligned} \tag{19}$$

By using that V' is increasing, we continue the estimate in (19) by

$$\sum_{k=j}^{n} V'(k) \geq \sum_{k=j}^{n} 4\int_{k-\frac{1}{4}}^{k} V'(r)\,dr \geq 4\int_{j-\frac{1}{4}}^{\infty} V'(r)\,dr = -4V\Big(j-\frac{1}{4}\Big),$$

and, similarly, $\sum_{j=1}^{n} V(j-\frac{1}{4}) \leq 4\int_{1/2}^{\infty} V = U(\frac{1}{2})$. We thus obtain that the left-hand side of (19) is bounded from below by $-16U(\frac{1}{2})$.

The first term in the right-hand side of (18) equals $E_{n-1}^1(x)$ with V replaced by V_n. Hence, (12) yields

$$\sum_{k=1}^{n-1}\sum_{j=0}^{n-1-k}\big[V_n\big((n-1)(x_{k+j}-x_j)\big) - V_n(k)\big] \gtrsim -[V_n''(2)]^{-1}\int_{\frac{1}{2}}^{\infty} U_n^2,$$

which, by a change of variables and $1 < \frac{n}{n-1} \leq 2$, we can further estimate from below by $-\Sigma/V''(2)$. In conclusion, we obtain from (18) and the estimates above that

$$V(\varepsilon_i^n + 1) \lesssim U(\tfrac{1}{2}) + \Sigma/V''(2) = \Lambda.$$

Since V^{-1} is decreasing, (15) follows.

Proof (*Theorem* 1) We rewrite

$$W_q^q(\varphi^n, \overline{\rho}) = \int_0^1 \big|\xi^n(t) - \overline{\xi}(t)\big|^q\,dt$$

$$
= \sum_{i=1}^{n} \int_{\frac{i-1}{n}}^{\frac{i}{n}} \left| \left(x_{i-1}^n + (t - \tfrac{i-1}{n}) n (x_i^n - x_{i-1}^n)\right) - (t - \tfrac{i-1}{n} + \tfrac{i-1}{n}) \right|^q dt
$$

$$
= \sum_{i=1}^{n} \int_{\frac{i-1}{n}}^{\frac{i}{n}} \left| (t - \tfrac{i-1}{n}) \varepsilon_i^n + \frac{1}{n} \sum_{j=1}^{i-1} \varepsilon_j^n \right|^q dt \leq \frac{1}{n} \sum_{i=1}^{n} \left| \frac{1}{n} \sum_{j=1}^{i} |\varepsilon_j^n| \right|^q \leq \frac{\|\varepsilon^n\|_1^q}{n^q} \tag{20}
$$

and

$$
\|\varphi^n - \overline{\rho}\|_{L^2(0,1)}^2 = \sum_{i=1}^{n} \int_{x_{i-1}^n}^{x_i^n} \left| \varphi^n(r) - 1 \right|^2 dr = \sum_{i=1}^{n} (x_i^n - x_{i-1}^n) \left| \frac{1}{n(x_i^n - x_{i-1}^n)} - 1 \right|^2
$$

$$
= \frac{1}{n} \sum_{i=1}^{n} \frac{1}{n(x_i^n - x_{i-1}^n)} \left| 1 - n(x_i^n - x_{i-1}^n) \right|^2 = \frac{1}{n} \sum_{i=1}^{n} \frac{(\varepsilon_i^n)^2}{\varepsilon_i^n + 1}. \tag{21}
$$

Then, Theorem 1 follows from Lemma 2. The inequality $\Theta^n \leq \frac{1}{n} \Xi$ follows from the fact that the infimum in (15) is smaller than or equal to 1 (because $\sum_{i=1}^n \varepsilon_i^n = 0$).

4 Convergence Rates in Different Scaling Regimes

In this section, we apply Theorem 1 for E_n as in (1) with general α_n. The goal is to express Θ^n and Ξ in terms of α_n, and to find sufficient bounds on α_n such that $\Theta^n \to 0$ and $\frac{1}{n} \Xi \to 0$ as $n \to \infty$.

For convenience, we write α instead of α_n whenever convenient. In this section, we set $V_\alpha := \alpha V(\alpha \cdot)$, and note that Theorem 1 implies

$$
W_q^2(\varphi^n, \overline{\rho}) \lesssim \Theta_\alpha^n := p\left(\frac{\Sigma_\alpha}{n V_\alpha''(2)^2} \right); \tag{22a}
$$

$$
\|\varphi^n - \overline{\rho}\|_{L^2(0,1)}^2 \lesssim \frac{\Xi_\alpha}{n} := \frac{1}{n} \frac{1}{V_\alpha^{-1}(\Lambda_\alpha)} p\left(\frac{\Sigma_\alpha}{V_\alpha''(2)^2} \right), \tag{22b}
$$

where Σ_α and Λ_α are defined by replacing V by V_α in (3d) and (8). Since for any $0 < A_1 < A_2$, the right-hand sides in (22) are bounded as a function of α on $[A_1, A_2]$, we find for any sequence (α_n) with $A_1 \leq \alpha_n \leq A_2$ for all $n \in \mathbb{N}$ that the convergence rates satisfy $\Theta_\alpha^n \lesssim \frac{1}{n}$ and $\frac{1}{n} \Xi_{\alpha_n} \lesssim \frac{1}{n}$. In fact, it holds that $\frac{1}{n} \Xi_{\alpha_n} \sim \frac{1}{n}$ if $V''(r)$ satisfies the same upper bound for $r \gg 1$ as in [7]. Indeed, it follows from [7, Theorem 4.3 and (7)] that ε^n converges in $\ell^2(\mathbb{N})$ to some $\varepsilon^\infty \not\equiv 0$, which implies by (21) the lower bound on the convergence rate.

To obtain estimates for the convergence rates when $\alpha_n \to \infty$ or $\alpha_n \to 0$ as $n \to \infty$, we need additional assumptions on V. We investigate these scenarios in the following two sections.

4.1 Convergence Rate as $\alpha_n \to \infty$

In addition to (3), we assume the existence of some $a > \frac{3}{2}$ such that

$$V(r) \lesssim r^{-a} \quad \text{and} \quad V''(r) \gtrsim r^{-(a+2)} \quad \text{for } r \gg 1. \tag{23}$$

The lower bound $a > \frac{3}{2}$ is required for (3d) to hold.

Next, we estimate Θ_α^n and Ξ_α for $\alpha \gg 1$. We obtain from $V_\alpha^{-1}(r) = \frac{1}{\alpha} V^{-1}(\frac{r}{\alpha})$, $V_\alpha^{(k)}(r) = \alpha^{k+1} V^{(k)}(\alpha r)$ and $U_\alpha(r) = U(\alpha r)$ that

$$\begin{aligned}
&V_\alpha''(2) = \alpha^3 V''(2\alpha) \gtrsim \alpha^{-(a-1)};\\
&U_\alpha(\tfrac{1}{2}) = U(\tfrac{\alpha}{2}) \sim \alpha^{-(a-1)};\\
&\Sigma_\alpha = \int_{\frac{1}{2}}^\infty U_\alpha(r)^2\,dr = \frac{1}{\alpha}\int_{\frac{\alpha}{2}}^\infty U(s)^2\,ds \sim \frac{1}{\alpha}\int_\alpha^\infty s^{-2(a-1)}\,ds \sim \alpha^{-2(a-1)};\\
&\Lambda_\alpha = U_\alpha(\tfrac{1}{2}) + \Sigma_\alpha / V_\alpha''(2) \lesssim \alpha^{-(a-1)} \xrightarrow{\alpha\to\infty} 0;\\
&V_\alpha^{-1}(\Lambda_\alpha) = \tfrac{1}{\alpha} V^{-1}(\tfrac{1}{\alpha}\Lambda_\alpha) \sim \tfrac{1}{\alpha}(\tfrac{1}{\alpha}\Lambda_\alpha)^{-\frac{1}{a}} \gtrsim 1;\\
&p\big(\Sigma_\alpha / V_\alpha''(2)^2\big) \lesssim p(1) \sim 1;\\
&\Theta_\alpha^n \lesssim \tfrac{1}{n}, \quad \Xi_\alpha \lesssim 1.
\end{aligned} \tag{24}$$

Hence, for any $\alpha_n \to \infty$ as $n \to \infty$, the convergence rates satisfy $\Theta_{\alpha_n}^n + \frac{1}{n}\Xi_{\alpha_n} \lesssim \frac{1}{n}$.

The estimates leading to (24) easily extend to the case in which (23) is relaxed to $V(r) \lesssim r^{-a}$ and $V''(r) \gtrsim r^{-(b+2)}$ for $r \gg 1$ and $b > a > \frac{3}{2}$. However, the resulting convergence rates will be larger, and may not converge to 0 if α_n is large enough with respect to n.

Another noteworthy test case is

$$V(r) = r \coth r - \log(2|\sinh r|), \tag{25}$$

which is considered in [3, 4, 8, 12] to model pile-ups of dislocation walls. It satisfies (3), $V''(r) \sim re^{-2r}$ for $r \gg 1$ and $V''(r) \sim r^{-2}$ for $r \ll 1$. The estimates leading to (24) result in Λ_α and Θ_α^n to grow exponentially fast in α, which, by the logarithmic singularity of V, leads to hyper-exponential growth of Ξ_α. Since we expect from [12] that the convergence rate should not be slower than the case in which $\alpha_n = 1$, the estimates in the proof of Theorem 1, in particular (10), are too coarse for V with exponential tails.

4.2 Convergence Rate as $\alpha_n \to 0$

In addition to (3), we assume existence of some $b > 0$ such that

$$V(r) \lesssim r^{-b} \quad \text{and} \quad V''(r) \gtrsim r^{-(b+2)} \quad \text{for } r \ll 1. \tag{26}$$

For $b \geq 1$, we note that $V \notin L^1(\mathbb{R})$, for which the author is not aware of any Γ-convergence result for E_n, or any convergence result of μ^n to $\overline{\rho}$, given that such a profile $\overline{\rho}$ exists. Nonetheless, a similar series of the estimates in (24) yields that for $b \neq \frac{3}{2}$[2]:

$$\begin{aligned}
&V_\alpha''(2) = \alpha^3 V''(2\alpha) \gtrsim \alpha^{1-b};\\
&U_\alpha(\tfrac{1}{2}) = U(\tfrac{\alpha}{2}) \sim \alpha^{1-b};\\
&\Sigma_\alpha = \int_{\frac{1}{2}}^\infty U_\alpha(r)^2\,dr = \frac{1}{\alpha}\int_{\frac{\alpha}{2}}^\infty U(s)^2\,ds \lesssim \frac{1}{\alpha}\Big[1 + \int_\alpha^1 s^{-2(b-1)}\,ds\Big] \sim \alpha^{-1-[3-2b]_-};\\
&\Lambda_\alpha = U_\alpha\Big(\frac{1}{2}\Big) + \frac{\Sigma_\alpha}{V_\alpha''(2)} \lesssim \begin{Bmatrix} \alpha^{-(2-b)} & b < \frac{3}{2} \\ \alpha^{-(b-1)} & b > \frac{3}{2} \end{Bmatrix} \xrightarrow{\alpha \to 0} \infty;\\
&V_\alpha^{-1}(\Lambda_\alpha) = \tfrac{1}{\alpha} V^{-1}(\tfrac{1}{\alpha}\Lambda_\alpha) \sim \tfrac{1}{\alpha}(\tfrac{1}{\alpha}\Lambda_\alpha)^{\frac{1}{b}} \gtrsim \alpha^{\frac{1}{b}[3-2b]_+};\\
&p\big(\Sigma_\alpha / V_\alpha''(2)^2\big) \lesssim p\big(\alpha^{-[3-2b]_+}\big) \sim \alpha^{-2[3-2b]_+};\\
&\Theta_\alpha^n \lesssim p(n^{-1}\alpha^{-[3-2b]_+}), \quad \Xi_\alpha \lesssim \alpha^{-(2+1/b)[3-2b]_+}.
\end{aligned} \tag{27}$$

For $b > \frac{3}{2}$, we obtain $\Theta_{\alpha_n}^n + \frac{1}{n}\Xi_{\alpha_n} \lesssim 1$ for any sequence $\alpha_n \to 0$. For $b < \frac{3}{2}$, we seek $\beta, \gamma > 0$ such that the lower bound $\alpha_n \gtrsim n^{-\beta}$ is enough to guarantee that the convergence rates

$$\Theta_{\alpha_n}^n \lesssim n^{-1+\beta[3-2b]_+}, \tag{28}$$

$$n^{-1}\Xi_{\alpha_n} \lesssim n^{-1+\gamma(2+1/b)[3-2b]_+} \tag{29}$$

converge to 0 as $n \to \infty$. For this convergence to hold, we require, respectively,

$$\beta < \frac{1}{3-2b} \quad \text{and} \quad \gamma < \frac{b}{(2b+1)(3-2b)}. \tag{30}$$

In conclusion, for $b > \frac{3}{2}$ and *any* $\alpha_n \to 0$, we obtain the new result that the minimisers μ^n of E_n converge to $\overline{\rho}$ (including convergence rates), even though there is no available Γ-convergence result on E_n. This new result extends to $0 < b \leq \frac{3}{2}$ whenever $\alpha_n \gtrsim n^{-\beta}$ or $\alpha_n \gtrsim n^{-\gamma}$ with β and γ as in (30). We further note that for $b < 1$ Γ-convergence of E_n is established in [12]. If, moreover, $\alpha_n \lesssim \frac{1}{n}$, then $\mu^n \rightharpoonup \tilde{\rho} \not\equiv \overline{\rho}$. This implies that β and γ cannot be larger than 1 for any $b < 1$, which is in line with the bounds in (30). In fact, the bound on β is sharp in the sense that it converges to 1 as $b \nearrow 1$.

From another point of view, [12] implies that $\mu^n \rightharpoonup \overline{\rho}$ for any $\frac{1}{n} \ll \alpha_n \ll 1$, while (29) requires a more restrictive lower bound on α_n for the convergence rates to converge to 0 as $n \to \infty$. We expect that this due to the nature of the estimates

[2]The case $b = \frac{3}{2}$ can be treated similarly; it makes $\log \alpha$ appear in the estimates.

from [7] which lead to Theorem 1. These estimates hinge on the intuition that each particle only interacts with a large, but n-independent, number of neighbours, and that they are closed to being equispaced.

Similar to Sect. 4.1, it is possible to relax (26) to $V(r) \lesssim r^{-a}$ and $V''(r) \gtrsim r^{-(b+2)}$ for $r \ll 1$ and $0 < b < a$. Also here, the resulting convergence rates will be larger.

Finally, we consider the specific potential defined in (25), which satisfies $V''(r) \sim r^{-2}$ as $r \ll 1$. The estimates leading to (27) result in $\Theta_\alpha^n \lesssim p(\frac{1}{n}\alpha^{-3})$ and exponential growth of Ξ_α in α. These results are consistent with (27) in the limit $b \to 0$.

5 Conclusion

Theorem 1 improves the results of [7] and [12] by guaranteeing convergence of the minimisers x^n of E_n to $\overline{\rho}$ in a stronger topology, together with a quantitative bound (the convergence rate). This is the first rigorous motivation for the remarkably good match between φ^n and $\overline{\rho}$ (see Fig. 2), which is observed in all numerical computations in [3, 4, 7] for similar energies as E_n in (1). Also, it is the first step for solving the open problem in [4] on choosing the best approximation $\overline{\rho}$ for fixed n and α_n.

The next challenge for solving this problem is to add an external potential to E_n. The complexity of adding such a potential can be seen from the fact that the minimiser $\overline{\rho}$ of the related limiting energy depends on the asymptotic behaviour of α_n [4]. Moreover, Theorem 1 relies on the compactness estimates in [7], which are not flexible for adding an external potential to E_n.

Another strategy for proving convergence rates in the presence of an external potential is to start from the techniques in [3, 4], which are designed to treat the singularity of V. This is work in progress which is intended for future publication.

References

1. Ambrosio, L., Gigli, N., Savaré, G.: Gradient Flows: In Metric Spaces and in the Space of Probability Measures. Birkhuser Verlag, New York (2008)
2. Brauchart, J.S., Hardin, D.P., Saff, E.B.: The next-order term for optimal Riesz and logarithmic energy asymptotics on the sphere. Contemp. Math. **578**, 31–61 (2012)
3. Garroni, A., van Meurs, P., Peletier, M.A., Scardia, L.: Boundary-layer analysis of a pile-up of walls of edge dislocations at a lock. Math. Models Methods Appl. Sci. **26**(14), 2735–2768 (2016)
4. Geers, M.G.D., Peerlings, R.H.J., Peletier, M.A., Scardia, L.: Asymptotic behaviour of a pile-up of infinite walls of edge dislocations. Arch. Ration. Mech. Anal. **209**, 495–539 (2013)
5. Hall, C.L.: Asymptotic analysis of a pile-up of regular edge dislocation walls. Mater. Sci. Eng. A **530**, 144–148 (2011)
6. Hall, C.L., Chapman, S.J., Ockendon, J.R.: Asymptotic analysis of a system of algebraic equations arising in dislocation theory. SIAM J. Appl. Math. **70**(7), 2729–2749 (2010)
7. Hall, C.L., Hudson, T., van Meurs, P.: Asymptotic analysis of boundary layers in a repulsive particle system. arXiv:1609.03236 (2016)

8. Roy, A., Peerlings, R.H.J., Geers, M.G.D., Kasyanyuk, Y.: Continuum modeling of dislocation interactions: why discreteness matters? Mater. Sci. Eng. A **486**, 653–661 (2008)
9. Saff, E.B., Kuijlaars, A.B.: Distributing many points on a sphere. Math. Intell. **19**(1), 5–11 (1997)
10. van Meurs, P.: Discrete-to-continuum limits of interacting dislocations. Ph.D. thesis, Eindhoven University of Technology (2015)
11. van Meurs, P., Muntean, A.: Upscaling of the dynamics of dislocation walls. Adv. Math. Sci. Appl. **24**(2), 401–414 (2014)
12. van Meurs, P., Muntean, A., Peletier, M.A.: Upscaling of dislocation walls in finite domains. Eur. J. Appl. Math. **25**(6), 749–781 (2014)

MIX
Papier aus verantwortungsvollen Quellen
Paper from responsible sources
FSC® C105338

If you have any concerns about our products,
you can contact us on
ProductSafety@springernature.com

In case Publisher is established outside the EU,
the EU authorized representative is:
Springer Nature Customer Service Center GmbH
Europaplatz 3, 69115 Heidelberg, Germany

Printed by Libri Plureos GmbH
in Hamburg, Germany